Edward Whymper

Supplementary appendix to travels amongst the great Andes of the equator

Edward Whymper

Supplementary appendix to travels amongst the great Andes of the equator

ISBN/EAN: 9783337157241

Hergestellt in Europa, USA, Kanada, Australien, Japan

Cover: Foto ©Andreas Hilbeck / pixelio.de

Weitere Bücher finden Sie auf **www.hansebooks.com**

SUPPLEMENTARY APPENDIX

TO

TRAVELS AMONGST THE GREAT ANDES

OF THE EQUATOR

BY

EDWARD WHYMPER

WITH CONTRIBUTIONS BY

H. W. BATES, F.R.S.	T. G. BONNEY, D.Sc., F.R.S.	G. A. BOULENGER.
PETER CAMERON.	F. DAY, C.I.E., F.L.S., F.Z.S.	W. L. DISTANT.
A. E. EATON, M.A.	F. D. GODMAN, F.R.S.	H. S. GORHAM, F.Z.S.
MARTIN JACOBY.	E. J. MIERS, F.L.S., F.Z.S.	A. SIDNEY OLLIFF.
O. SALVIN, F.R.S.	DAVID SHARP, M.B., F.R.S.	T. R. R. STEBBING, M.A.

ILLUSTRATED

Quæ fuit durum pati,
Meminisse dulce est.
SENECA

LONDON
JOHN MURRAY, ALBEMARLE STREET
1891

PREFACE.

I HAVE already explained in the narration of my Travels amongst the Great Andes of the Equator the circumstances under which the collections were formed that are described in this Supplementary Appendix to that work. It is said there, and it may be desirable to repeat here, that, whilst it was my aim to secure all we might discover in the highest zones of the Andes of Ecuador, it was not intended to attempt to examine zoologically the lower regions of that country. The latter have often been worked by professional collectors, and they are easy of access, and can be explored with comparatively little trouble at any time. The loftier and highest regions, on the other hand, had not previously been examined; they offer nothing that is attractive to a commercial collector; they are more or less difficult of access, and they are, in consequence of the violent and rapid meteorological disturbances which frequently occur, well-nigh inaccessible to all except those who are prepared to remain for a length of time at a great height, provided with such equipments as will afford efficient protection against the inclemency of the weather.

After our experiences in these elevated regions, I should call the work of collection in them one of the most arduous that a Botanist or Zoologist could undertake. In the intervals of fine weather acquisitions are likely to be small, and there are long periods in which nothing can be accomplished; and, since my return, I have felt even more strongly than before our departure, it is improbable that for a considerable length of time any one will be delegated to, or will impose upon himself, the task of collecting either in Zoology or Botany amongst the *highest* zones of the Andes. For this reason, as well as for the others which have been given elsewhere, it seemed of more importance and utility to give such moments as could be spared from our other work to research in the higher and highest zones than to attempt to investigate the lower regions.

We accordingly pushed rapidly across the lower country both in going to and returning from the interior;[1] and only acquired, whilst passing through it, such objects as came readily to hand. In respect to the interior, it should be noted (1) that the exigencies of travel often caused us to traverse considerable stretches of country without attempting to collect at all; (2) that, at such places as we stopped at, our researches never even *approached* an exhaustive character; and (3) that the more minute species were rejected, owing to the known difficulty of inducing specialists to undertake their examination. Bearing these various points in mind, it appears to me improbable that in the interior (say, in the areas more elevated than 8000 feet) we obtained as much as *one-tenth* of the number of species which might have been collected by a person who could have given his whole time and attention to zoological research.

Amongst the Insects collected from the level of the sea up to 8000 feet, 16 per cent are new to science. One hundred and sixty species were obtained from 8000 feet and upwards, and of these exactly 60 per cent were previously unknown; and at the greatest heights (15-16,000 feet) the whole are new. The following table exhibits

[1] In speaking of the "interior," it is to be understood that I refer to tracts of country seldom less elevated than 8500 feet. The neighbourhood of the town of Ibarra (7300 feet), and the bottom of the great ravine of Guallabamba (6472 feet), were the only localities we visited in the interior which were at a lower level. During the 212 days we passed in the interior, there were only four upon which we were at a lower elevation than 6000 feet.

It has not been considered necessary to place the word 'Ecuador' after the *habitats* which are quoted throughout this volume; but it should be understood that the whole of the localities which are mentioned are situated in that country.

PREFACE.

the increasing proportion of unknown to known species the higher we ascended:—

Height.	No. of species obtained.	Previously known.	New to science.	Not identified.
8- 9,000 feet	16	12	4	...
9-10,000 ,,	48	21	22	5
10-11,000 ,,	3	...	3	...
11-12,000 ,,	34	4	24	6
12-13,000 ,,	18	3	13	2
13-14,000 ,,	17	4	11	2
14-15,000 ,,	14	3	10	1
15-16,000 ,,	9		9	...
above 16,000 ,,	1	...		1
Totals	160	47	96	17

It may possibly be inferred, from the comparatively large number of beetles which were secured, that the Coleoptera much preponderate over other orders of Insects upon the Great Andes of the Equator. Such an opinion would, I think, be erroneous, though at the very greatest heights they are possibly as numerous as all other insects put together. Yet various Diptera range almost as high as the highest of the Coleoptera, and I can count up 75 species of spiders which were obtained at 9500 feet and upwards. At moderate elevations in Ecuador —say 10-11,000 feet—spiders were apparently more numerous than anything else, and at some localities, such, for example, as Machachi, they swarmed in countless numbers. Few Hymenoptera were found anywhere near the snow-line, and of this order it may be remarked that some of the largest species are found towards the superior limit of its range, which is an exception to the general rule.[1]

[1] I strongly dissent from the statement made by the late Prof. James Orton (in "Contributions to the Natural History of the Valley of Quito," in the *American*

The insects referred to in this volume number 359 species, and of these less than one-half come from 8000 feet and upwards. In the remaining (undescribed) collections the proportion of species obtained from these altitudes is much larger, and there are probably amongst them not less than 350 species which were collected by myself and my assistants at heights greater than 9000 feet. It is with very great regret I find myself compelled, after so long delay, to publish only a portion of the results which were obtained on the journey, and unable to present a more substantial contribution towards the Zoology of the Great Andes of the Equator. Should it be found practicable, the remaining results may one day appear as a Second Supplement.

COLEOPTERA. The beetles which were obtained at heights greater than 15,000 feet belong to the *Pterostichinæ*, *Otiorrhynchidæ*, and *Curculionidæ*, and are described by Mr. H. W. Bates and Mr. S. Olliff. The strongly-marked characters of the species in the latter groups rendered the selection of specimens more easy than in the genera dealt with by Mr. Bates; and, partly owing to this, comparatively few examples came into Mr. Olliff's hands.[1] In most instances, the beetles which were obtained at the greatest heights were discovered by tearing up roots, digging, or by turning over stones, and I do not recall a single occasion on which they were found actually upon the surface. The two *Colpodes* which are described as *C. megacephalus* (p. 13) and *C. Pichinchæ* (p. 15), both came from the highest peak of Pichincha,—the

Naturalist, Nov. 1872, p. 651) that insects are few in number in the interior. In localities with vegetation, they are often conspicuously numerous, and they are not wanting in the most arid districts. But, in saying this, I mean that they are conspicuously numerous to those who will search for them. At the *greatest* altitudes scarcely anything was obtained except by diligent searching; and, as the majority of the insects which are obtainable are dull in coloration and small in size, they may readily be overlooked. Of the more minute insects (say those less than four millimètres in length), myriads can be found at heights between 9-13,000 feet.

[1] The smallness of my means, and the necessity of keeping down baggage to the lowest possible point, frequently caused the rejection of many duplicates which it would perhaps have been advisable to have retained.

PREFACE. ix

former from the very highest point, and the latter from a place about 600 feet below, at which we encamped. In each case the insects were discovered in course of breaking out rock specimens, and were disinterred from amongst stones which were cemented together with ice. It is scarcely exaggeration to say that they were imbedded amongst the stones. In these instances, and in many others, the *Colpodes* were found in groups or clusters. The *Curculios*, on the contrary, were commonly met with as isolated individuals.

The most widely-diffused beetle that we observed in the interior is the *Astylus* described by Mr. Gorham (p. 52). It was collected at various heights between 9-13,500 feet, and was found almost everywhere within that range, congregated in such numbers that hundreds of specimens might have been obtained from a single bush. As this insect is of moderate size, and readily catches the eye, it is surprising that it has not been described long since. The beetle *Leucopelaea albescens*, Bates (p. 30), is also a remarkable example of oversight. This was found, in the first instance, upon a sandy plain to the northwest of Cotopaxi. Vegetation was scanty at this spot, and the insects, which were in large numbers, quickly attracted the eye. Dead as well as living were spread over the ground for a distance of several miles; and, although about a dozen only were secured, many hundreds might have been taken. This region has been traversed by several, at least, of my predecessors. It is indeed obvious that the middle zones of Ecuador have been very imperfectly worked by collectors, even in the localities most frequently visited by them, such as Quito. At the southern outskirts of that city there is a prominent hill called the Panecillo,[1] which is now almost surrounded by houses, and is used as a playground by the youth of the place. I visited this eminence one day alone, to obtain a round of angles, and by beating the dwarf vegetation into my hat secured about thirty species of insects of various orders, without any expectation that a place so frequented

---

[1] See the plan of Quito accompanying the narrative. The summit of the Panecillo is almost exactly 10,000 feet above the level of the sea.

would yield interesting results. Yet amongst the few which have been described there are two new genera, and nothing that was collected appears to have been obtained before.[1]

FORMICIDÆ. A very small number of Ants were met with in the interior. All were insignificant in size, and at no place were they so numerous as to be an inconvenience,—though the reverse was the case at Guayaquil. It was at this place, in my bedroom, I secured, casually, the only remarkable ant obtained on the journey—the *Holcoponera* described by Mr. Cameron (p. 92).

LEPIDOPTERA. A great number of Butterflies are found in the interior, belonging to a comparatively small number of species. The lower zones, on the other hand, are exceedingly prolific in species; and it is no exaggeration to say that a larger number may be obtained in some of them in one *day* than can be secured from 8000 feet upwards in an entire *year*.[2] In returning towards Guayaquil we took what is termed the railway route, and were arrested at the Bridge of Chimbo (about 1000 feet above the sea) by want of a train. Whilst waiting, in little more than half an hour, we collected the twenty-three species which will be found enumerated between pp. 96-110, on a piece of ground not more than 300 yards long by 100 broad, and saw not less than a dozen others which we should have secured had we worked a little longer. In the interior, from 8000 feet upwards, our endeavours during six months only procured 28 species, and we certainly did not see so many as half a dozen others which were not captured.

[1] The following appear in this volume. *Plectonotum nigrum*, Gorham (pp. 51-2); *Listrus cacseeas*, Gorham (p. 53); *Seymnus?* (p. 58); *Apion Andinum*, Ollif' (p. 78); *Luperosoma marginata*, Jacoby (p. 87); and *Pheidole monticola*, Cameron (p. 93). The Panecillo had been visited by Humboldt and Bonpland, Buckley, Ida Pfeiffer, Reiss and Stübel, and by many others.

[2] I am informed by Messrs. Godman and Salvin that the described species of Lepidoptera which are reputed to have come from Ecuador now exceed one thousand; and there is little doubt that the lower zones contain many more as yet unknown. Very incorrect localities have been, I think, quoted as habitats of the described species.

By far the most abundant butterfly in the interior of Ecuador is the *Colias* which has been identified by Messrs. Godman and Salvin as *C. dimera* (p. 108); and excluding one other (*Pieris xanthodice*) its individuals possibly exceed in number the individuals of all the other species put together. It was collected at numerous localities between 7200-13,000 feet, and was seen almost everywhere within that range. It was especially abundant along the banks of streams, and was often found congregated in hundreds over muddy spots or marshy soil. The much less abundant *Colias* which has been identified as *C. lesbia* fluttered in company with its more numerous relatives, and upon several occasions the two were taken at one sweep of the net.

The highest flying butterfly, and (with the exception of three or four beetles) the highest insect of any kind obtained, barring stragglers, is the *Colias* which is described by Messrs. Godman and Salvin as *C. alticola* (p. 107). I have elsewhere identified it as the same species which was seen upon Chimborazo by Humboldt and Bonpland.[1] This was actually collected between 12,000-16,000 feet, and was observed slightly higher. Its highest range therefore exceeds the height of the mean snow-line.[2] It is scarcely possible that we can have overlooked other species of diurnal Lepidoptera at the greatest altitudes; and, in the absence of others, *Colias alticola* must be regarded as the highest-flying butterfly in either of the two Americas.

In number of individuals, *Pieris xanthodice* (p. 106) is inferior only to *Colias dimera*, and in the height which it attains it stands second to *Colias alticola*, but its *range in altitude* is greater than that of either. It was observed in localities somewhat below 9000 feet up to a little higher than 15,000 feet, and at all intermediate points. Its range in altitude therefore exceeds 6000 feet, which is a larger amount than was observed in the case of any other butterfly. It was moderately abundant over the whole of the country we traversed, and was

[1] See *Travels amongst the Great Andes*, Chap. XIX., and the accompanying figure.
[2] The snow-line in Ecuador, as in other countries, varies upon different mountains, and upon different sides of the same mountain. In no place should I be disposed to regard it as higher than 16,000 feet.

conspicuously so upon the great slopes of the basin of Machachi, where it was frequently met with, flying in large companies, in rather open order.[1]

RHYNCHOTA. A small insect that has been referred to the genus *Emesa* (p. 117) was the solitary example of its order which was taken at a great elevation. This was obtained at 16,500 feet on the southern side of Illiniza (about a thousand feet above the snow-line), and was obviously a straggler. We saw no other living thing at so great a height as this, and the only animal remains which were obtained at an equal or greater altitude were the partly fossilized bones which were found at 18,000 feet on the southern side of Chimborazo.[2]

The interior of Ecuador is rich in this order of insects, and the small extent of our collection is due to the fact that we were unable to give the necessary time for research in the most favoured localities.

CRUSTACEA. It is observed in Chapter V. of *Travels amongst the Great Andes* that Crustacea appear to be scarce in the interior of Ecuador. Five species only were obtained,[3] and these, although new to the country, were all previously known. The three species of Woodlice were both numerous and widely distributed, and it is not easy to understand how they have been overlooked by others. One of these, namely *Metoponorthus pruinosus*, is amongst the exceptional species with a great range in altitude, having been taken with my own hands on the banks of the Guayas at the level of the sea, at the Hacienda of Antisana (13,300 feet) and at several intermediate points. The single Amphipod, *Hyalella inermis*, S. I. Smith (p. 125), captured at the Hacienda of Antisana, in the Valley of Collanes (12,500 feet), and near Machachi (9800 feet), was numerous at each of those localities, and is probably widely distributed in the interior. According to the Rev. T. R. R.

[1] Other remarks upon the diurnal Lepidoptera are more conveniently given in the volume of narrative. A figure of *Pieris xanthodice* is given in Chap. XIX.
[2] *Travels amongst the Great Andes*, Chap. III.
[3] *Pseudothelphusa macropa*, S. I. Smith (p. 121); *Philoscia angustata*, Nicolet (p. 125); *Porcellio lævis*, Latreille (p. 125); *Metoponorthus pruinosus*, Brandt (p. 125); and *Hyalella inermis*, S. I. Smith (p. 125).

Stebbing, to whom I am indebted for its identification, no Amphipod has hitherto been recorded from so considerable an elevation.[1]

REPTILIA. Three species of Lizards were obtained in the interior, which have been identified by Mr. Boulenger as *Liocephalus trachycephalus* (A. Dum.); *Ecpleopus* (*Pholidobolus*) *montium*, Peters; and *Proctoporus unicolor* (Gray). The *Liocephalus* was numerous throughout the interior generally, between elevations of a little over 8000 feet to a little under 12,000 feet, and was more frequently noticed in the northern than in the southern part of the country,—being especially abundant on the Plain of Tumbaco (to the N.N.E. of Quito), less frequently seen in the basin of Machachi, while to the south of the basin of Riobamba it only occurred occasionally. *Ecpleopus montium* was less numerous, yet still was far from being rare, and was rather widely distributed; but the *Proctoporus* (p. 130) was only obtained on the eastern side of the Plain of Tumbaco, and in the contiguous basin of Chillo. This little lizard (averaging only 4½ inches in length), when caught, turned upon his captors, bit fiercely, and could hang on with its jaws to a finger or anything which was presented for it to snap at. There are probably two other species of Lizards, inhabiting the interior, which we failed to capture.

In the higher and highest parts of the interior we neither saw Snakes nor could learn of the existence of any. The two solitary specimens which were obtained both came from the lowest basins, and were brought in alive by natives.[2] The examples of *Bothrops Schlegeli* were presented to me by Mons. Giacometti, *maître d'hôtel*, at Quito, who obtained them upon his farm; which, according to his description, is placed a considerable distance to the west of Quito, and is probably situated at a low altitude, but I was unable to procure any precise information about it. This snake is said to be very particularly venomous, and is greatly feared.

[1] A figure of *Hyalella inermis* is given in *Travels amongst the Great Andes*, Chap. XIX.
[2] These were brought in through rewards being offered.

BATRACHIA. Only four species of Frogs were collected in the interior, namely, *Phryniscus lævis*, Gthr.; *Hylodes unistrigatus*, Gthr.; *H. Whymperi*, Bouleng.; and *Nototrema marsupiatum* (Dum. & Bibr.). Of these, the first mentioned is I think the most widely distributed, and the last named is the most numerous. In the vicinity of the town of Machachi we saw thousands, and must have heard hundreds of thousands of frogs,[1] principally of this latter species,—which is very variable in its colouring and markings. Of the *Hylodes* it may be remarked that we obtained all our specimens upon *the ground*. The species with which Mr. Boulenger has associated my name was observed only at rather highly-placed localities, and seems very restricted in its range in altitude; whereas *Phryniscus lævis* is seen at almost every height between 7000-13,500 feet.

PISCES. The readers of Humboldt's works will remember the remarkable statements which were made by him about the little siluroid fish which he described and figured as *Pimelodus cyclopum*.[2] I should scarcely have been led to make any search or enquiry for this fish if I had not seen the remarks by Dr. Putnam in the *American Naturalist*, 1871, p. 694, and learnt that Humboldt's fish appeared to have been described upon five or six different occasions, under as many different names. Dr. Putnam advanced the opinion that the whole of these so-called different species should be referred to one, somewhat variable, species. The descriptions were based either upon single specimens, or upon a very small number of examples, and I thought it advisable to procure a considerable number, from different localities, so that the whole subject might be re-investigated. Several hundreds were procured on the spot; these were reduced to fifty-one, and I had the advantage of submitting them, upon my return, to the independent examination of Dr. F. Day, who coincides with the views expressed by Dr. Putnam.[3]

---

[1] Compare this with Orton; "of frogs there are not enough to get up a choir." *The Andes and the Amazons*, English ed., p. 107.

[2] *Observations de Zoologie et d'Anatomie comparée*, vol. i. pp. 21-5, pl. 7, Paris, 1811; and in *Aspects of Nature*, vol. ii. p. 231.

[3] Some remarks upon this fish will be found in my Chapter upon Cayambe.

PREFACE.

It is now my duty to acknowledge the assistance which has been so kindly rendered by the eminent specialists who have examined these collections;[1] and, whilst thanking all, I should very especially thank my old friends Mr. H. W. Bates and Professor T. G. Bonney. Besides the important contributions from the pen of Mr. Bates, I am greatly indebted to him for having acted throughout as my entomological adviser. The extensive series of rocks which was brought home has been carefully and thoroughly investigated by Professor Bonney, but it is thought sufficient to present in this volume only a few general remarks, as his observations have already been published in full in the Proceedings of the Royal Society.

Amongst my friends in Ecuador who have rendered assistance, I should particularly mention and thank Mr. G. Chambers, British Consul at Guayaquil, for the snakes which he presented from that locality. At Chillo, I received valuable aid from a sharp little English lad, Master Willie Slater, who volunteered to collect for me in a district which I was obliged to leave too soon. His diligence was rewarded by obtaining several of the species which are here first described. In Quito, help was rendered by a highly-intelligent Swede (to whom I had been recommended by Baron de Thielmann), Ludwig by name, and through him I acquired the greater part of the collections which come from Milligalli, Tanti, and other places in the West of Ecuador—a region which there was no time to visit.

Lastly, my most sincere thanks are due to my two assistants, Jean-Antoine and Louis Carrel, for their zeal and industry upon all occasions. Our altitudes were determined by mercurial barometers, and the laborious duty of transporting these instruments devolved upon J.-A. Carrel. In consequence of his extreme care, no breakages occurred; and the heights of the localities which are mentioned throughout this volume are accordingly fixed with a degree of

---

[1] The authors are solely responsible for their respective contributions. The whole of the Type Specimens of the Insects remain in the possession of the authors of the papers; and those of the Reptiles and Frogs have been acquired by the British (Natural History) Museum.

accuracy which could not have been attained by any other method that could have been employed.[1] Careful observations for altitude were of the first importance in this country. Within its various zones, it contains almost as many climates, and as great a range of temperature as the entire continent; and, if the collections had no better habitat than "Ecuador" attached to them, scarcely more information would be afforded than if they were said to have come from South America. Through the hearty co-operation of my two assistants, something was obtained from every locality we visited, and every specimen was accurately labelled and catalogued.[2]

In his *Vues des Cordillères*, Humboldt deplores the small results which have been attained upon high mountain expeditions in the following passage: "Ces excursions pénibles, dont les récits excitent généralement l'intérêt du public, n'offrent qu'un très-petit nombre de résultats utiles au progrès des sciences." This statement has been substantially true, and it has conveyed a reproach alike to the men of science who have *not* investigated the loftier portions of the earth's surface, and to those who *have* penetrated them without making use of their opportunities. It has commonly been taken for granted that the tracts in the neighbourhood of the snow-line, or rising above it, are either lifeless or are denizened by stragglers. Could the whole of our acquisitions have been presented here, they would have demonstrated that the upper zones of Ecuador—even the tracts closely bordering the snow-line—are far from being lifeless. They would have exhibited forms the most extraordinary, of wondrous diversity; they would have defined the upper range of many species; and would, doubtless, have supplied various missing links. I repeat the expres-

---

[1] Tables of altitudes and temperatures are given in *Travels amongst the Great Andes*.

[2] In course of setting the insects, various specimens from Machachi (9800 feet) were mixed with others from the Pacific slopes (7-8000 feet). It was not possible to identify the whole from recollection, and I destroyed those which could not be referred to their proper localities, with the exception of the five species described at pp. 25, 26, 40, 50 and 68.

sions of my deep regret that it is only possible to present some fragments of our results. Yet, enough I trust appears to encourage my contemporaries in mountain-travel to continue similar researches, laborious and unthankful though they may be; gradually to amass such a body of evidence as will in course of time render no longer true the dictum of my illustrious predecessor; and will permit it to be said, instead, that high-mountain explorations, although perhaps of little interest to the general public, are of great value to Science.

EDWARD WHYMPER.

CONTENTS.

	PAGES
INTRODUCTION, by H. W. BATES, F.R.S.	1–6
COLEOPTERA, by H. W. BATES, F.R.S.	7–39

    ,,   Fam. CICINDELIDÆ.
    ,,   ,,   CARABIDÆ.
    ,,   ,,   COPRIDÆ.
    ,,   ,,   TROGIDÆ.
    ,,   ,,   MELOLONTHIDÆ.
    ,,   ,,   RUTELIDÆ.
    ,,   ,,   DYNASTIDÆ.
    ,,   ,,   CETONIIDÆ.
    ,,   ,,   PASSALIDÆ.
    ,,   ,,   PRIONIDÆ.
    ,,   ,,   CERAMBYCIDÆ.
    ,,   ,,   LAMIIDÆ.

Do. *(continued)*, by DAVID SHARP, M.B.	40–44

    ,,   Fam. DYTISCIDÆ.
    ,,   ,,   SILPHIDÆ.
    ,,   ,,   STAPHYLINIDÆ.
    ,,   ,,   TENEBRIONIDÆ.

Do. *(continued)*, by the Rev. HENRY S. GORHAM, F.Z.S.	44–58

    ,,   Fam. ELATERIDÆ.
    ,,   ,,   DASCILLIDÆ.
    ,,   ,,   LYCIDÆ.
    ,,   ,,   LAMPYRIDÆ.
    ,,   ,,   TELEPHORIDÆ.
    ,,   ,,   MELYRIDÆ.
    ,,   ,,   PTINIDÆ.
    ,,   ,,   HISPIDÆ.
    ,,   ,,   CASSIDIDÆ.
    ,,   ,,   EROTYLIDÆ.
    ,,   ,,   COCCINELLIDÆ.

		PAGES
COLEOPTERA (continued), by A. SIDNEY OLLIFF		58–81
,,	Fam. NITIDULIDÆ.	
,,	,, TROGOSITIDÆ.	
,,	,, OTIORRHYNCHIDÆ.	
,,	,, CURCULIONIDÆ.	
,,	,, CALANDRIDÆ.	
,,	,, BRENTHIDÆ.	
Do. (continued), by MARTIN JACOBY		82–88
,,	Fam. EUMOLPIDÆ.	
,,	,, CHRYSOMELIDÆ.	
,,	,, HALTICIDÆ.	
,,	,, GALERUCIDÆ.	

HYMENOPTERA (FORMICIDÆ), by PETER CAMERON . . . 89–95

LEPIDOPTERA, by F. DUCANE GODMAN, F.R.S., & OSBERT SALVIN, F.R.S. 96–110

RHYNCHOTA, by W. L. DISTANT 111–120

CRUSTACEA (PODOPHTHALMIA), by EDW. J. MIERS, F.L.S., F.Z.S. . 121–124
    Do. (ISOPODA), by the Rev. A. E. EATON, M.A. . . . 125
    Do. (AMPHIPODA), by the Rev. T. R. R. STEBBING, M.A. . 125–127

REPTILIA AND BATRACHIA, by G. A. BOULENGER . . . 128–136

CYCLOPIUM CYCLOPUM, HUMBOLDT, by F. DAY, C.I.E., F.L.S., F.Z.S. 137–139

NOTE ON ROCKS FROM THE ANDES, by Prof. T. G. BONNEY, D.Sc., F.R.S. 140–143

INDEX TO GENERA 145–147

LIST OF ILLUSTRATIONS.

The Figures have been drawn by R. J. COOMBS, W. HERBERT, W. PURKISS, E. WILSON, and others; and have been engraved on wood by EDWARD WHYMPER.

PLATES.

1.	LEUCOPELÆA ALBESCENS, BATES	To face Page	31
	BAROTHEUS ANDINUS, BATES	,,	31
2.	BARYXENUS ÆQUATORIUS, BATES	,,	32
3.	HETEROGOMPHUS WHYMPERI, BATES	,,	33
4.	PRAOGOLOFA UNICOLOR, BATES	,,	34
5.	PRIONOCALUS WHYMPERI, BATES	,,	36
6.	HAMMODERUS STICTICUS, BATES	,,	39
7.	STRONGYLIUM DENTICOLLE, SHARP	,,	42
	ASIOPUS OPATROIDES, SHARP	,,	42
8.	PLECTONOTUM NIGRUM, GORHAM	,,	51
	SILIS CHIMBORAZONA, GORHAM	,,	51
9.	HELICORRHYNCHUS VULSUS, OLLIFF	,,	60
	PLEURONECES MONTANUS, OLLIFF	,,	60
10.	DIBOLIA VIRIDIS, JACOBY	,,	84
	LUPEROSOMA MARGINATA, JACOBY	,,	84
	DORYPHORA PICTURATA, JACOBY		84
11.	HOLCOPONERA WHYMPERI, CAMERON	,,	92
12.	LYDE TRANSLUCIDA, DISTANT	,,	113
	NEOMIRIS PRÆCELSUS, DISTANT		113
	DIONYZA VARIEGATA, DISTANT		113
13.	SQUILLA DUBIA(?), M. EDWARDS		124
14.	CYCLOPIUM CYCLOPUM, HUMBOLDT	,,	137

FIGURES IN THE TEXT.

		PAGE
1.	TRACHYDERES VERMICULATUS, BATES	6
2.	ANISOTARSUS BRADYTOIDES, BATES	8
3.	PTEROSTICHUS (AGRAPHODERUS) ANTISANÆ, BATES	10
4.	COLPODES ALTARENSIS, BATES	16

LIST OF ILLUSTRATIONS.

	PAGE
5. COLPODES STENO, BATES	20
6. UROXYS LATESULCATUS, BATES	24
7. ONTHERUS ÆQUATORIUS, BATES	25
8. CLŒOTUS TUBERICAUDA, BATES	26
9. CLAVIPALPUS ANTISANÆ, BATES	27
10. PLATYCŒLIA PRASINA, ERICHS.	29
11. CYCLOCEPHALA RUBESCENS, BATES	31
12. GYMNETIS FLAVOCINCTA, BATES	35
13. EURYSTHEA ANGUSTICOLLIS, BATES	38
14. CARNEADES NODICORNIS, BATES	39
15. ATHOUS DISPAR, GORHAM	45
16. PLATEROS? ALTICOLA, GORHAM	47
17. CLADODES NIGRICOLLIS, GORHAM	47
18. XENISMUS WHYMPERI, GORHAM	50
19. ASTYLUS BIS-SEXGUTTATUS, GORHAM	53
20. COMPSUS WHYMPERI, OLLIFF	64
21. EXORIDES CARINATUS, PASCOE	65
22. LISTRODERES PUNCTATISSIMUS, OLLIFF	70
23. MACROPS CŒLORUM, OLLIFF	72
24. ANCHONUS MONTICOLA, OLLIFF	73
25. HILIPUS LONGICOLLIS, OLLIFF	75
26. OTIDOCEPHALUS? SPINICOLLIS, OLLIFF	77
27. BRENTHUS VULNERATUS, GYLH.	81
28. CAMPONOTUS MAYRI, CAMERON	90
29. PHEIDOLE MONTICOLA, CAMERON	94
30. CINYPHUS? OBSCURUS, DISTANT	115
31. STENOPODA SCUTELLATA, DISTANT	116
32. PNOHIRMUS WHYMPERI, DISTANT	117
33. ACANTHIA ANDENSIS, DISTANT	118
34. CARINETA FIMBRIATA, WALKER	119
35. TETTIGONIA DUPLICARIA, DISTANT	120
36. PSEUDOTHELPHUSA MACROPA, M. EDWARDS	122
37. PSEUDOTHELPHUSA MACROPA, VAR. PLANA, S. I. SMITH (?)	122
38. CORONELLA WHYMPERI, BOULENGER	131
39. PROSTHERAPIS WHYMPERI, BOULENGER	133
40. PHRYNISCUS ELEGANS, BOULENGER	134
41. HYLODES WHYMPERI, BOULENGER	135
42. HEAD AND VENTRAL FIN OF CYCLOPIUM CYCLOPUM, HUMBOLDT	138

ADDENDA.

The two following species have been found by Mr. H. W. Bates since the sheets of his paper were worked off. — E. W.

20*. *Colpodes quadricollis*, Chaudoir, Ann. Soc. Ent. Fr., 1859, p. 300; id. ibid. 1878, p. 287.

Hab. Corredor Machai (12,779 feet).

40*. *Bembidium Andinum*, n. sp.

Hab. Chimborazo, west side (15,811 feet). Eight examples.

B. fulvocincto affine; minor et minus convexum, fulvescenti-æneum, subauratum politum, subtus cum pedibus castaneo-rufum.

Long. $3\frac{1}{2}$ millim.

The head is scarcely narrowed behind the eyes, which are only very moderately prominent, and the elytra are ovate with the slight humeral angles close to the hind angles of the thorax. The frontal furrows are broad and shallow, and between them and the eye respectively the surface is flat, with the anterior setiferous pore large and conspicuous. The thorax is rather more narrowly cordate than in *B. fulvocinctum*, strongly sinuated and narrowed behind the middle, and with acute hind angles; the basal fovea large and deep, and the side near the angle not distinctly carinated. The elytra are obsoletely striated, the striæ being visible only when viewed obliquely and from the apex; the two marginal striæ being closely approximated in a marginal groove. Antennæ and palpi dusky, scape reddish.

SUPPLEMENTARY APPENDIX

INTRODUCTION.

By H. W. Bates, F.R.S.

The total number of species of Insecta and Arachnida collected by Mr. Whymper during his journey in Ecuador amounts to about one thousand. Unfortunately it has been found impossible to induce specialists to work up the whole of the groups for the purpose of the present volume; several important families and whole orders remain unnamed, and are therefore for the present unavailable in aiding us to form some idea of the nature and relations of the Fauna of the Equatorial Andes. A rough estimate has been made of the numbers of species in the missing groups. Thus Baron von Osten Sacken, on looking over the Diptera, considered them to number about 100; Mr. Druce, who partially determined the Moths (*Lepidoptera Heterocera*), found 44 species; the Hymenoptera (exclusive of the Ants) appear to be scarcely less numerous than the Diptera, and the Spiders comprise not fewer than 200 species.

The Orders and Families of the Insecta class enumerated or described in this Supplementary Appendix comprise 359 species. Of these no fewer than 131 are new to science, and many of them are so distinct that 14 new genera have had to be instituted for their reception.

So much interest attaches to the nature of the Insect Fauna of high altitudes in the Equatorial zone of the Andes, and to its relations to the Faunas of Chili and Temperate zones of North America and Europe, that it would be undesirable to let the occasion pass of analysing Mr. Whymper's collection with this view, notwithstanding that so large a portion must be left out of the examination. The few remarks I have to make must further be restricted almost exclusively to the Coleoptera. There is the less disadvantage in this that the collections made by Mr. Whymper at very high altitudes consisted largely of this order of Insects, and that Coleoptera, by their great number and ubiquity, and the tendency to definiteness in their areas of distribution, offer better material for this class of inquiry than almost any other group of organisms.

The remarkable relationship which exists between the Fauna and Flora of Chili and those of high latitudes in North America, and even Europe, has often been discussed by writers on Geographical Distribution. In Darwin's *Origin of Species* it is stated (on the authority of Sir Joseph Hooker) that not fewer than 46 species of flowering plants of Chili and Europe are identical,—North America showing evidence of its having lain in the path of the supposed migration. The number of flowering plants common to these regions and the remarkable general similarity in their Botanical products has been, quite recently, shown by Dr. Philippi, the Chilian Botanist, to be much greater even than stated by Darwin. An analogous relationship has also been pointed out in their animal forms. In Insects, for example, numerous genera are common to the three regions, which are totally absent from the intervening Tropical and warm Temperate zones of America. The curious and puzzling feature in these classes of facts is that, as far as Insects and Plants are concerned, the relationship is not of the same degree. In Plants there is a large amount of identity as to species, but in Insects there is little or none; the relationship is generic only. The explanation of the proportional identity of Plants offered by Darwin—namely, that the species migrated along the high lands of the Andes from north to south during the Glacial epoch, does not meet the case of the more distant relationship of the respective Insect Faunas. Darwin's explanation had been previously applied to account for similar relationship between the Faunas of south and north temperate latitudes in the old world, and has been almost universally adopted. With good reason, for here we have the remarkable evidence of its validity offered by the products of high altitudes within the old world tropics, and those of low lands near the Arctic zone and the mountains of temperate latitudes. Multitudes of closely-allied or identical species are now found in these various localities, thus indicating the paths that species took, when driven by the great climatic changes of the Glacial period. The question then arises, Have we any similar proof of a glacial migration in Tropical America?

Any final conclusion on this point, at any rate as regards the Insects, has hitherto been deferred, on account of our ignorance of the products of the high Andes at elevations near the snow-line. The researches of Humboldt and Bonpland in this direction were unsatisfactory, as no species were obtained at great elevations; and, like those famous travellers, subsequent explorers have brought away from the upper slopes and valleys of Ecuador and Colombia, until within the last few years, only Tropical American forms, or species of Andean genera closely allied to the products of the plains at their feet. There was no trace of the great host of genera and species characteristic of similar local conditions in north and south temperate latitudes. The mountain-living genera *Carabus* and *Nebria* of the old world temperate fauna and of North America, or their Chilian allies *Ceroglossus*, *Migadops* and very many others, certainly ought to be found, if the inferences drawn from the botanical relationship were as sound as they appear to be. Shortly before Mr. Whymper's wonderful ascents of the snow-capped peaks of Ecuador, a few Coleoptera and Lepidoptera from high elevations there and in Colombia had reached Europe, and went far to discourage any expectations; but we may say that Mr. Whymper's more thorough and complete search, at all elevations and on nearly all the loftier mountains, has set the question at rest. If there had been any distinct element of a North Temperate or South Temperate Coleopterous Fauna on the Ecuadorian Andes the collections he made, inexhaustive though they may be, would have shown some traces of it; but there are none. A few genera belonging to temperate latitudes, though not found in the tropical lowlands, do indeed occur, but they are forms of almost world-wide distribution in similar climates, and there is no representative of the numerous characteristic and common genera of the north or south. Even the northern genera more or less abundantly found on the Mexican highlands are absent.

The species of Coleoptera collected at altitudes above 9000 feet are about 100 in number. The majority of these are mountain representatives of genera characteristic of the lower levels or the plains of Tropical America, some of which genera (*e.g. Colpodes, Silpha, Philonthus*, etc.) occur also in other parts of the world; others (*e.g. Anisotarsus, Pelmatellus, Platycoelia, Leucopelea, Trigonogenius*, etc.) are American genera, having Chilian or Antarctic affinities, whilst others, as far as we at present know, are peculiar to the high Andes. One feature of the Fauna is of great interest. It is the occurrence of apterous species of genera which at lower levels are always winged; of these Mr. Whymper found two belonging to the genus *Bembidium*, and one, a weevil, belonging to the genus *Macrops*. This apterous condition has been dwelt upon by Darwin, as a significant characteristic of the Coleoptera of lofty mountains and Oceanic Islands. All the species of *Bembidium* found on the island of St. Helena are apterous, forming a distinct group within the genus.

The butterflies enumerated in this Appendix are nearly all species of the Tropical American lowlands, or closely allied to them. The exceptions are the few which were found flying at elevations from 10,000 to 16,000 feet, viz. *Pieris Xanthodice, Colias alticola*, and *Colias dimera*, all three belonging to small groups of their respective genera which occur at great altitudes throughout South America, and as far south as Chili. The genera *Erebia, Chionobas, Parnassius, Argynnis, Epinephele* and many others, so highly characteristic of the Faunas of the North Temperate zone, or Chili, or both, and of high vertical ranges, are quite absent.

It seems to me a fair deduction from the facts here set forth that no distinct traces of a migration during the lifetime of existing species, from north to south, or *vice versa*, along the Andes, have as yet been discovered, or are now likely to be discovered. It does not follow, however, that the Darwinian explanation of the peculiar distribution of species and genera on mountains in the Tropical and Temperate zones, and in high latitudes of the old world, is an erroneous one. The different state of things in the new world is probably due to the existence of some obstacle to free migration, as far as regards Insects, between north and south, both during and since the Glacial epoch. The problem, like most others relating to Geographical Distribution, is a complicated one; but there are one or two considerations, likely to be overlooked, which may tend to its solution. One is the great altitude at which the vigorous denizens of the teeming tropical lowlands flourish on the slopes of the Andes. Mr. Whymper found, for example, species of many of the genera of Longicorn Coleoptera characteristic of the lowland forests at altitudes of 9000 and 10,000 feet, and Kirsch has recorded numerous species of *Lampyridæ*, *Lycidæ*, and other families belonging equally to Tropical American forest genera, as met with by Reiss and Stübel in Colombia and Ecuador at 12,000 feet. In Ecuador all the warm moisture brought by the Eastern trade-winds is not intercepted even now by the wall of the Andes, and wherever that falls, in the depressions, conditions of climate and vegetation will be created suitable to these encroaching Tropical forms. If we add to this the barrenness and generally unfavourable conditions of the zone above those altitudes, there can be little wonder that Temperate forms have not freely passed along the Andes. Another consideration is that there may have been a breach of continuity of the land in Glacial times, at the Isthmus of Panama, sufficient to prevent free migration. It may, further, be legitimate to speculate on the possibility of the Andes being lower in the Tropical zone during the Glacial epoch. A few hundred feet lower than the present altitude, combined with the copious warm rains which must have accompanied the age of ice, would present conditions undoubtedly favourable to the spread of Tropical forms over the whole area, which would successfully resist the invasion of high northern or southern species. The main principle in distribution, however, is that forms sooner or later, and in proportion to their intrinsic and extrinsic facilities of dissemination, will find their way all over the world to wherever the conditions inorganic and organic are favourable to their acquiring a footing. That these facilities are possessed in a higher degree by Plants than Insects and some other groups of animals may be a sufficient explanation of the fact that so many species of Plants have surmounted the obstacles to their passage from north to south during the last Glacial epoch, while few or no Insects have done so. The more distant, or generic, relationship between the Insects of Chili and those of the North Temperate zone can only be explained on the assumption of a migration at some epoch far more remote than the last Glacial epoch.

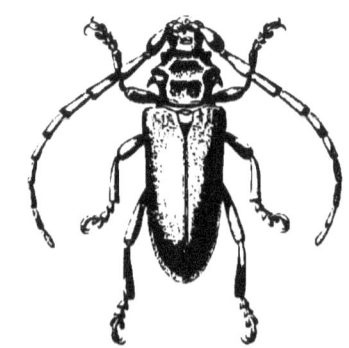

TRACHYDERES VERMICULATUS, BATES.
NEAR CHILLO.

COLEOPTERA.[1]

By H. W. BATES, F.R.S., F.L.S.

Tribe ADEPHAGA.

Fam. CICINDELIDÆ.

1. *Pseudoxycheila bipustulata*, Latreille, Voy. de Humboldt & Bonpland, Ins. p. 153, t. 16, figs. 1, 2 (*Cicindela*).

Hab. Nanegal (3-4000 feet). A single example. Latreille gives the banks of the Amazons as the locality of Humboldt and Bonpland's specimens. In his description he seems to include this and the next form (*P. angustata*) in one species.

2. *P. angustata*, Chaudoir, Cat. Coll. Cicindelidæ, p. 62.

Hab. Milligalli (6230 feet [2]). I have specimens from North Peru, and from Macas in Ecuador taken by Buckley, and, according to Chaudoir, it has also been obtained in East Peru. Mr. Whymper's specimen is much narrower in form, both of thorax and elytra, than *P. bipustulata*, and generally is of a greenish instead of a blue colour.

Fam. CARABIDÆ.

Subfam. ANISODACTYLINÆ.

3. *Anisotarsus Peruvianus*, Dejean, Spec. Gen. Col., iv, p. 289.

Hab. Nanegal (3-4000 feet). One example, doubtfully referred to this species, the thorax being more quadrate, i.e. with straighter sides and less rounded hind angles. Dejean's type-specimens came from S. Lorenzo, in Peru; they vary in colour from bright green or blue to bronzed black, and the ♂, if not also the ♀, is shining on the upper surface.

---

[1] The whole of the Coleoptera have been described under the general direction of Mr. H. W. Bates. His personal contribution extends from p. 7 to p. 39.—*E. W.*

[2] The altitude of Milligalli is given on the authority of Father Menten. See Proc. Royal Geog. Soc., p. 489, Aug. 1881.—*E. W.*

4. *A. Bradytoïdes*, n. sp.

Hab. Penipe to Riobamba (9000 feet); Machachi (9-10,000 feet); Illiniza (14,000 feet). Numerous examples.

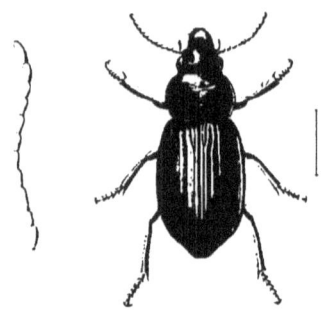

ANISOTARSUS BRADYTOIDES, BATES.
ILLINIZA, MACHACHI, ETC.

A. Peruviano proxime affinis ; differt colore suprà semper obscure fusco-cupreo (raro viridi-tincto) elytris ♂ ♀ subtiliter alutaceis, sericeo-nitentibus, tarsis sicut tibiis femoribusque piceis nec rufo-fulvis ; corpore breviori et latiori ; elytrorum interstitiis planissimis.

Long. 8-10 millim. ♂ ♀.

Subfam. PELMATELLINÆ.

5. *Pelmatellus rariipes*, n. sp.

Hab. Machachi (9-10,000 feet); between Latacunga and Machachi (9200 feet); Quito (9350 feet); Pichincha (12,000 feet); Hacienda of Guachala (9217 feet); Pacific slopes (7-8000 feet). Numerous examples.

P. nitescenti (Bates) simillimus, sed differt thoracis angulis posticis omnino rotundatis, femoribusque (interdum tibiisque apice) nigro-fuscis.

Long. $5\frac{1}{2}$-$6\frac{1}{2}$ millim. ♂ ♀.

Difficult to be distinguished from the common and widely-distributed Central American *P. nitescens* (Bates, in Salvin and Godman, Biologia Centrali-Americana. Col., vol. i, p. 68, pl. iii, fig. 17), but the blackish femora and the more rounded hind angles of the thorax are conspicuous and pretty constant differential characters ; the species varies in the colour of the legs, the apex of the tibiae and the tarsi being also sometimes more or less blackish, the basal part of the tibiae only remaining pale reddish. *P. obtusus* (Bates, ejusd.

op.) has rounded hind angles to the thorax, but also red legs, and is not so glossy in its metallic coloration as *P. nitescens* and *P. rarüipes*. The form of the thorax varies slightly in all the species of the genus.

6. *P. oxynodes*, n. sp.

Hab. Machachi (9-10,000 feet); Ibarra (7000 feet); Quito (9350 feet); Valley of Collanes, Altar (12,500 feet); Pichincha (12,000 feet). Numerous examples.

Longior elytris elongatis, postice subdilatatus, nigro-æneus politus, antennis basi (cæteris piceis) palpisque rufis, pedibus nigro vel rufo-piceis tibiis basi pallidioribus: thorace transversim quadrato ante basin leviter sinuato, angulis posticis exstantibus, rectis (interdum acutis) elytris subtiliter punctulato-striatis, interstitiis planissimis, tertio impunctato.

Long. 6-7 millim. ♂ ♀.

Similar to *P. nitescens* in the outline of the thorax, but the hind angles much more salient, never subobtuse but sometimes acute and always preceded by a distinct sinuation of the lateral margins, which also are a little flattened out at the angle. The legs are usually pitchy, nearly black, with base of tibiæ paler, but they vary to reddish. In all the varieties, however, the species may be distinguished from *P. nitescens* by the absence of the usual discoidal puncture on the third interstice, as also by the elongate elytra, the very much finer striæ and darker bronzed black colour.

7. *P. Andium*, n. sp.

Hab. Chuquipoquio, Chimborazo (12-13,000 feet); Tortorillas, Chimborazo (13,300 feet); Cotocachi (11-13,500 feet). Very numerous examples.

Elongato-ovatus, castaneo-rufus, æneo-tinctus vel toto-fusco-æneus, antennis basi palpis pedibusque rufis: thorace rotundato transverso, lateribus regulariter arcuatis: elytris apice late obtusis vix sinuatis, fortiter striatis interstitiis subplanis tertio impunctato.

Long. 7-7½ millim. ♂ ♀.

Distinguished by its very rounded thorax, in which there is no trace of hind angles. The narrow marginal rim is continued uniformly round the angles towards the scutellum, and the usual basal fovea is sublinear, very faintly impressed and impunctate. The third elytral interstice has no trace of puncture in the numerous specimens I have examined, and the scutellar striole is also wanting.

8. *P. Guachalensis*, n. sp.

Hab. Hacienda of Guachala (9217 feet). Two examples.

P. Andium simillimus, sed differt thoracis fovea basali profunda, punctata, elytrorumque striola scutellari punctoque interstitii tertii munitis. Minor,

castaneo-rufus, æneo-tinctus capite obscuriori, politissimus: thorace rotundato, elytris acute striatis interstitiis planissimis.
Long. 5½ millim. ♀.

9. *Pelmatellus* ——— ?
Hab. Forests above the Bridge of Chimbo (1-3000 feet). Two specimens, females, possibly of this genus, but not determinable from this sex alone.

Subfam. POLPOCHILINÆ.

10. *Polpochila scaritides*, Perty, Del. Anim. Artic. Bras. Ins., p. 13, t. 3, f. 7.
Hab. Guayaquil (indoors). Two examples. Has also been found on the Upper Amazons.

Subfam. PTEROSTICHINÆ.

11. *Pterostichus* (*Agraphoderus*) *Antisanæ*, n. sp.
Hab. Hacienda of Antisana (13,300 feet); La Dormida, Cayambe (11,800 feet); southern side of Chimborazo (12-13,000 feet). Seven examples.

Omaseo minori (Gyll.) haud dissimilis. Oblongus, piceo-niger nitidus (♂♀) antennis palpis pedibusque piceo-rufis: capite lævi sulcis frontalibus sat tenuibus: thorace late quadrato, lateribus sat rotundatis prope basin sinuatis angulis posticis rectis, fovea basali utrinque (a margine longe distanti) sublineari vix impressa: elytris oblongis basi rectis humeris brevissime dentatis, apice profunde sinuatis, striatis, interstitiis planis tertio bipunctato puncto 1mo prope basin ad striam tertiam (interdum absenti) 2do paullo post medium prope striam secundam.

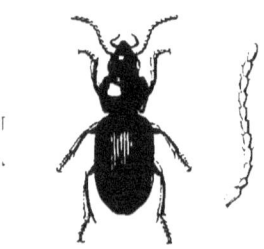

PTEROSTICHUS (AGRAPH.) ANTISANÆ, BATES.
ANTISANA, CHIMBORAZO, ETC.

Segmento ultimo ventrali simplice, utrinque ♂ unipunctato ♀ bipunctato.
Long. 7½-8½ millim.

Similar in form to a small *Omaseus* (e.g. O. minor Gyll.), but rather broader. The marginal stria of the elytra is single, the prosternal process faintly margined, the palpi tapering at the apex and briefly truncated, the ventral segments simple with a pair of setiferous punctures in the middle.

12. *Pt.* (*Agraphoderus*) *Pichinchæ*, n. sp.
Hab. Pichincha (12,000 feet).

Pt. Antisanæ proxime affinis, differt solum statura minori, thoraceque lateribus omnino rotundatis, angulis posticis obtusissimis.
Long. 7 millim. ♂.

One example only, differing from *Pt. Antisanæ* almost solely in the form of the thorax, its sides being equally and moderately rounded and behind curving obliquely, without the slightest trace of sinuation, to the base, forming a very obtuse angle: the base on each side has a smooth and rounded fovea, slightly impressed midway between the median line and the angle. The elytra are of the same form, the striæ strongly marked and the interstices plane, but the anterior puncture on the third is absent. As in *Pt. Antisanæ* there is a rather short scutellar striole obliquely placed between the first and second striæ.

13. *Pt. (Agraphoderus) liodes*, n. sp.

Hab. Hacienda of Antisana (13,300 feet), Machachi (9-10,000 feet). Three examples.

Pt. Antisanæ affinis sed major et latior. Oblongus, brevis, niger ♂ nitidus, ♀ elytris subopacis, palpis antennis pedibusque piceo-rufis: thorace relative magno quadrato lateribus æqualiter parum arcuatis, angulis posticis subrectis (apice obtusis) foveis basalibus utrinque duabus vix impressis: elytris mox pone basin paullo ampliatis deinde sat parallelis, apice sinuatis, striatis, interstitiis planis tertio bipunctato (puncto 1mo apud medium 2do longe ante apicem), interdum tripunctato.

Long. 9 millim. ♂ ♀.

Belongs to the same small group of the genus as the two preceding, but is larger and relatively broader; the thorax not sinuated before the hind angles as in *Pt. Antisanæ*, but less rounded than in *Pt. Pichinchæ*, the sides forming with the base an angle a little more open than a right angle; there is a faint indication of a second basal fovea, near the angle. The elytra differ in being rather less oblong (a little more rounded behind the humeral angle) and in having two punctures on the third interstice placed quite differently from those in *Pt. Antisanæ*, the first being about the middle and the second midway between the middle and the apex. In one of the three examples there is a third anterior puncture, towards the base.

14. *Pt. (Agraphoderus) integer*, n. sp.

Hab. Southern and eastern sides of Chimborazo, from Tortorillas to Chuquipoquio (13,300 to 11,700 feet). Very numerous examples.

Elongato-ovatus, niger nitidus, antennis palpis pedibusque piceo-rufis: thorace relative magno, antice angustato, lateribus modice arcuatis, basi lato angulis posticis subrectis (apice valde rotundatis) foveis basalibus plerumque obsoletis: elytris mox pone basin paullulum ampliatis deinde usque ad apicem angustatis, sat profunde striatis interstitiis planis vel subplanis tertio punctis 2-4 (interdum quinto bipunctata); abdomen subtus ♂ ♀ sicut in *Pt. Antisanæ*.

Long. 8-10½ millim. ♂ ♀.

Distinguished from its allies by its peculiar outline, the thorax being narrowed (in some examples much narrowed) anteriorly and the elytra slightly tapering posteriorly from behind the shoulders. In other respects it differs very little from *Pt. lioctes*. The discoidal punctures of the elytra appear normally to be three in number, one in the middle, one midway between that and the base, and the third midway between the middle one and the apex, but one or other of them is wanting in some examples, or supplementary punctures appear, in one case even two on the fifth interstice.

The four preceding species all agree in the fusiform apical joint of the palpi, moderately impressed frontal furrows, marginel prosternum, short metathoracic episterna, and simple apical ventral segment in the ♂, and they differ from the allied genera *Pterostichus*, *Haptoderus*, etc., sufficiently to require their separation as a distinct group. I propose for the group the name *Agraphoderus*.

15. Pterostichus ———?

Hab. Between Machachi and Pedregal (10,000 feet).

A single immature and distorted ♀ example, possibly belonging to the restricted genus *Pterostichus*, having deep thoracic foveæ, simple apex of prosternum, etc.

16. *Anchomenus Quitensis*, n. sp.

Hab. Quito (9350 feet). A single example.

A. *Chilensi* (Dej.) affinis, sed thorace angustiori, angulis posticis distinctis. Gracilis, subæneo-niger, politus, palpis antennis pedibusque piceis ; capite angusto, sulcis frontalibus profundis, oculis magnis : thorace anguste cordato, postice sat angustato sed haud sinuato angulis posticis obtusis (apice acutis) margine basali prope angulum utrinque obliquo : elytris oblongis, apice obtusis, plica basali cum margine laterali ad humeros nullomodo angulum formanti, tenuiter striatis, striis 5-7 subobsoletis, interstitiis planis tertio 3-punctato.

Long. 7 millim.

Belongs to a group of species inhabiting temperate South America, of which A. *Chilensis* may be cited as the type. The fourth tarsal joint in the anterior tarsi is emarginated, but in the middle and posterior pair is rather narrow and triangular ; the tarsi are rather slender, finely setose beneath and sulcated on the sides of the basal joint in the four hinder legs ; the metathoracic episterna are long and narrow.

17. *A. (Sericoda) decempunctatus*, Reiche, Rev. Zool., 1843, p. 310 (*Dromius*).

Hab. Cayambe village (9320 feet), and between Machachi and Pedregal (10,000 feet). Two examples.

A widely-distributed species at high elevations in Colombia, Chiriqui, Guatemala, Mexico.

18. *A. (Agonum) Andicola*, n. sp.

Hab. Eastern slopes of Pichincha (12,000 feet). One example.

Sat breviter ovatus, nigro-æneus, nitidus, scapo tibiis femoribusque subtus obscure rufis: capite ovato, lævi, linea tenui impressa a fovea frontali usque ad oculum ducta: thorace breviter ovato, postice magisquam antice angustato, angulis posticis nullis, margine laterali usque ad medium basin paullo elevato: elytris postice sat perspicue sinuatis, acute striatis striis 5-7 subobsoletis, interstitiis planis tertio 4-punctato.

Long. 5½ millim.

Broader and relatively shorter than the European *A. fuliginosus*, the thorax shorter and more circular, the colour a dark brassy, the striæ fine but sharply impressed, except 5-7 which are very faint and the interstices perfectly plane.

19. *Colpodes megacephalus*, n. sp.

Hab. On the summit ridge of Guagua Pichincha (15,600 feet); Hacienda of Antisana (13,300 feet); and Cayambe (12-14,000 feet). Seven examples.

Oblongus, piceo-niger, ♂ nitidus ♀ alutaceus subopacus, palpis antennis pedibus pectoreque (plus minusve) castaneo-rufis: antennis in hoc genere brevibus: capite magno pone oculos elongato-tumido, collo crasso haud transversim impresso: thorace quadrato postice gradatim subrecte angustato, angulis posticis subobtusis: elytris subtilissime punctulato-striatis interstitiis planissimis tertio 4-7 punctato: tarsis subtus dense ciliatis articulo 4to emarginato.

Long. 13-15 millim. ♂ ♀.

Belongs to the small group of species of which *C. cephalotes* (Chaud.) is a member, peculiar to high elevations in the Andes, and distinguished by the very large head, elongated anteriorly and tumid around and behind the eyes, the rather short antennæ and obtusely-rounded apex of the elytra. It is much broader and larger than *C. cephalotes*, and resembles, at first sight, a *Pterostichus* or small *Percus*, but the head has only very shallow frontal furrows. The palpi are moderately elongated, with the apical joints nearly cylindrical. The thorax has advanced anterior angles and obtuse posterior angles rounded at their apices, the sides being nearly straight without the slightest sinuation near the base. The soles of the tarsi have a rather dense brush of short moderately stiff hairs, and the sides of the basal joints have only faint longitudinal furrows; the penultimate joint is strongly emarginate, almost bilobed in the anterior pair, and having the outer angles somewhat elongated in the hindmost pair; the claw joint has a few cilia underneath. The metathoracic episterna are short, the outer side about equal in length to the breadth at the base. The elytra are oblong, very slightly narrowing to the nearly rectangular shoulders; the punctures on the third interstice are more than three in number, sometimes as many as seven, the two basal ones being on the third stria or in the middle of the third interstice—the others near the second stria.

20. *C. capito*, n. sp.

Hab. Altar (12,500 feet). A single example.

Elongatus, elytris elongato-ovatus, nitidus, (♀) castaneo-fuscus, partibus oris antennis pedibusque castaneo-rufis: capite sat magno, oculis prominentibus, post oculos recte angustato, collo supra transversim depresso: thorace quadrato postice subrecte angustato, margine laterali postice valde reflexo angulis rectis: elytris apice conjunctim subacuminatis vix perspicue sinuatis, punctulato-striatis, interstitiis planis, tertio tripunctato: tarsis articulo 4^(to) bilobato, posticis lobo exteriori elongato.

Long. 11 millim. ♀.

Belongs to the *C. cephalotes* group by the comparatively short antennæ with joints more ovate than in the typical *Colpodes*. The metathoracic episterna are short, but the outer side appears a little longer than the width at the base. The head differs much in shape from *C. megacephalus*, having a constricted neck as usual in the genus, and the sides, instead of forming an elongate tumour behind the eyes, narrowing obliquely and almost in a straight line to the neck, the eyes standing out very prominent. The soles of the tarsi are rather densely clothed with short bristles, the claw joint is naked.

21. *C. seriepunctatus*, Chaudoir, Ann. Soc. Ent. Fr., 1859, p. 298.

Hab. High plateaux of Colombia. Mr. Whymper took a specimen that answers fairly well to the description at La Dormida, Cayambe (11,800 feet).

22. *C. pustulosus*, n. sp.

Hab. Cayambe (15,000 feet). One example.

Oblongo-ovatus castaneo-rufus: capite ovato antice cum mandibulis brevi collo crasso supra nullomodo depresso, oculis minime convexis: antennis brevibus articulis 5-10 quadrato-ovatis: thorace breviter quadrato postice valde sinuato, angulis posticis exstantibus acutis, supra prope angulum utrinque elevatione conspicua rotunda pustuliforme: elytris ovatis apice obtuse conjunctim rotundatis, basi utrinque depressis plica basali obsoleta, striis fere obsoletis, sed interstitiis convexis tertio unipunctato: tarsis 4 anticis articulo 4to profunde, 2 posticis vix, emarginatis.

Long. 6½ millim. ♀.

This species has a peculiar facies owing to the shortness of the mandibles, and the consequent bluntness of the muzzle, so different from the usual aspect of *Colpodes*. The maxillæ project far beyond the mandibles, and the apical joint of the palpi is fusiform. The thorax is similar in form to that of many *Pterostichi*, but presents the singular feature of a rounded bladder-like elevation near each posterior angle,—the metathoracic episterna are very short.

23. *C. rotundiceps*, n. sp.

Hab. Cayambe (15,000 feet). One example.

C. pustuloso affinissimus, sed multo gracilior, castaneo-fuscus capite nigro, partibus oris antennis pedibusque rufioribus: antennis sat gracilibus: capite ovato antice cum mandibulis brevi collo transversim depresso, oculis minime convexis: thorace elongato cordato-quadrato, postice valde sinuato fere constricto, angulis exstantibus acutis, spatio inter foveam et angulum elongato-convexo: elytris ovatis, apice obtuse rotundatis, basi utrinque depresso sed plica basali integra, subtiliter punctulato-striatis, 3^{io} quinque-punctato: tarsis 4 anticis haud profunde, 2 posticis vix perspicue, emarginatis.

Long. 9½ millim. ♀.

Very similar to *C. pustulosus*, but longer and slenderer in all its parts; differing specifically, besides, in the presence of a well-marked basal plica on the elytra. The mandibles in *C. pustulosus* project very little beyond the labrum, scarcely the labrum's length; in *C. rotundiceps* they project about 1½ the length of the labrum; the maxillæ project far beyond the mandibles in both species. The usual tumid enlargement behind the eyes is elongated, but gradually narrows to the neck which has a faint constriction. The thorax is very much narrower and resembles in form that of *C. piceolus* (Chaudoir).

24. *C. Pichinchæ*, n. sp.

Hab. Pichincha, second camp (15,000 feet). Three examples.

C. pustuloso proxime affinis, differt capite antice mandibulisque longioribus fere sicut in *Colpodes* normalibus, collo angustiori supra perparum depresso, oculis prominentibus capite post oculos tumido rotundatim angustato. Castaneo-rufus; thorace subcordato-quadrato, postice valde sinuato, angulis exstantibus subacutis, inter foveam et angulum convexo: elytris apice late obtusis, basi utrinque depressis plicaque obsoleta, subtilissime striatis, interstitio tertio 3-punctato, puncto primo prope basin alteris duobus longe post medium.

Long. 7½ millim. ♂ ♀.

Closely allied to *C. pustulosus*, having similar short antennae, feebly emarginated posterior tarsi, depressed base of each elytron, etc.; but differing in the shape of the head and the more oblong elytra. The thorax is similar, but narrower and more elongated, and the rounded pustule-like elevation near the posterior angle is reduced to a simple convexity, without distinct rounded outline. The claw joint of the tarsi has beneath a few very short bristles.

25. *C. orthomus*, Chaudoir, Ann. Soc. Ent. Fr., 1878, p. 289.

Hab. Hacienda of Antisana (13,300 feet), Cayambe (12-15,000 feet). Ten examples.

In Chaudoir's description, drawn from a single specimen, the sides of the thorax are said to be "presque droits"; they are nearly straight in some examples but distinctly sinuated in others, the general form however being always trapezoidal, as Chaudoir indicates. The thorax is relatively very large, and the elytra short in this species, which resembles a short *Calathus* rather than a *Colpodes*. Chaudoir speaks of the scutellar striole as effaced, a term not strictly applicable as there exists a short rudiment. The base of the thorax between the smooth fovea and the angle has a convexity resembling the more pronounced pustular elevation of *C. pustulosus*.

26. *C. Altarensis*, n. sp.

Hab. Valley of Collanes, Altar (12,500 feet). Four examples.

Angustus, niger vix nitidus; antennis brevibus; capite parvo, collo perparum constricto: thorace subovato, postice paullo angustato, angulis posticis rotundatis, nullis: elytris apice minus late rotundatis nec sinuatis ♂ ♀ alutaceis, subtiliter striatis interstitio tertio 3-punctato: tarsis subtus sparsim setosis articulo 4to modice emarginato.

Long. 7½-8½ millim. ♂ ♀.

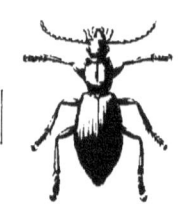
COLPODES ALTARENSIS, BATES. ALTAR.

Resembles certain slender species of the genus *Agonum*. The head is normal, with rather prominent eyes, and neck scarcely constricted. The terminal joint of the palpi is fusiform. The legs and antennae are relatively short and stout. The metathoracic episterna are very short.

27. *C. denigratus*, n. sp.
Hab. Pichincha (12,000 feet). One example.

C. Altarensi simillimus sed differt antennis elongatis: gracilior, niger elytris (δ) alutaceis subopacis, antennis partibusque oris fulvo-rufis; capite parvo, collo haud transversim depresso, oculis parum prominentibus; thorace sicut in *C. Altarensi* quadrato-ovato angulis posticis nullis, marginibus angustis; elytris longioribus, anguste oblongis, apice perparum sinuatis, plica basali fere recta, transversa, humeris rotundatis, punctulato-striatis, interstitiis planissimis tertio tripunctato; tarsis plantis sat dense tenui-setosis 4to quatuor posticorum lobo exteriori conspicue elongato.

Long. 9 millim. δ.

Very similar to *C. Altarensis* in form, sculpture, and colour, but different in the greater length of the antennæ, and the elongation of the outer angle of the fourth tarsal joint to the middle and hind legs, which assumes the proportions of a lobe; so that the joint may be said to be unilobular. The insect is of more oblong form, but without the structural differences in antennæ and tarsi it would not be considered as more than a local variety of *C. Altarensis*.

28. *C. fusipalpis*, n. sp.
Hab. Cayambe (15,000 feet). One example.

Sat gracilis, niger, antennis palpis pedibusque piceo-rufis: antennis brevibus sed articulis minus ovatis; capite ovato, oculis parum prominentibus, collo transversim modice depresso; palpis articulo ultimo late fusiformi; thorace angusto quadrato ante basin profunde sinuato, angulis posticis exstantibus fere rectis; elytris ovatis subtiliter striatis interstitio tertio tripunctato; tarsis articulo 4to sat profunde emarginato, posteriorum angulis productis.

Long. 6½ millim.

Apparently allied to *C. oopteroïdes* (Chaud.), and possibly the same species, but Chaudoir does not mention the thick fusiform terminal joint of the maxillary and labial palpi; the swollen part of the joint being much wider than the penultimate. The antennæ and legs are short and stout, the thorax is narrow, quadrate but deeply sinuated before the base, and the hind angles salient and sharp at their apices; the base is rugulose and there is a slight linear convexity between the fovea and the angle. The elytra are elongate-ovate, the base narrow with well-marked curved basal plica, and the apex rounded without sinuation. The metathoracic episterna are remarkably short, and the fourth tarsal joint is much more deeply emarginated (nearly bilobed) than in *C. pustulosus* and allies.

29. *C. Patroboïdes,* n. sp.

Hab. Valley of Collanes, Altar (12,500 feet). One example.

C. capito affinis et similis, sed multo gracilior capiteque minus dilatato, etc.; piceo-niger antennis palpis pedibusque castaneo-rufis: capite normali oculis prominentibus, collo constricto: thorace quadrato postice paullo angustato ante basin sinuato angulis posticis rectis subacutis, margine laterali prope angulos anticos explanato, postice nec explanato nec elevato: elytris elongato-ovatis convexis (♂) nitidis, apice perparum sinuatis, punctulato-striatis: tarsis subtus sparsim setosis, articulo 4to emarginato, posteriorum angulis breviter elongatis.

Long. 8 millim. ♂.

Allied to *C. fusipalpis*, but larger and more slender in all its parts. In the form of the thorax and elytra (particularly the dilated anterior margins of the thorax), and in the prominent eyes with the sides of the head converging obliquely and rapidly from the eyes to the somewhat constricted neck, the species resembles also *C. capito*, but it differs greatly from that species in the clothing of the soles of the tarsi and the feebler emargination of their fourth joint, besides its smaller head and the narrower, unreflexed posterior margins of the thorax.

30. *C. orcas,* n. sp.

Hab. Western side of Chimborazo, fifth camp (15,800 feet). One example.

Elongatus, nigro-piceus, palpis antennis pedibusque castaneo-rufis: antennis modice elongatis: capite ovato, collo transversim depresso, post oculos parum convexos paullulum tumido, rotundatim angustato: thorace subcordato, postice perparum sinuato angulis posticis rectis nec productis: elytris valde elongatis, basi angustis apice obtuse rotundatis, subtiliter punctulato-striatis, interstitiis planissimis tertio 4-punctato: tarsis subtus pauci-setosis articulo 4to profunde emarginato fere bilobato, posticis lobo exteriori elongato.

Long. 10½ millim. ♂.

Elongate in form, with oval head, the eyes being scarcely more prominent than the tumid cheeks behind them, which latter narrow in a slightly curved line to the moderately constricted neck. The palpi have slender terminal joints, a little wider in their middle portion. The antennae are a little more slender than in *C. cephalotes*. The thorax is cordate with rectangular hind angles and much reflexed but narrow lateral margins, which are widest at the anterior angles. The elytra are relatively very long but somewhat ovate in outline. The claw joints of the tarsi are naked beneath. The metathoracie episterna are rather narrow, the outer margin being a little longer than the base.

31. *C. laevilateris,* n. sp.

Hab. Between Tortorillas and Chuquipoquio, Chimborazo (12-13,000 feet). Two examples.

Elongatus, gracilis, castaneo-fuscus nitidus (♂) antennis palpis pedibusque rufioribus : capite antice cum mandibulis elongato, oculis prominentibus post oculos recte angustato, collo angusto sed supra parum depresso ; antennis sat elongatis ; thorace relative parvo, cordato-quadrato postice paullo sinuato angulis rectis sed apice haud acuminatis ; elytris elongato-ovatis basi angustis apice paullo sinuatis subobsolete striatis, interstitiis interioribus convexis exterioribus planis politis tertio 5-punctato ; tarsis omnibus articulo 4to bilobato lobo exteriori vix longiori.

Long. 9 millim. ♂.

A slender and elongated species with relatively long but not large head elongated mandibles, prominent eyes, and narrow elongate-ovate elytra, the 4-7th striae of which are very faintly impressed. The legs are slender, the tarsi clothed beneath rather densely with moderately fine hairs, and the fourth joint in all bilobed, the outer lobes a little longer and broader than the inner. The metathoracic episterna are very short. The thorax differs from that of *C. piccolus* and allied species in not being deeply sinuated behind or having produced posterior angles, the lateral margin being straight for some distance before the angle. The colour of the thorax is a little redder than that of the head and elytra.

This species is evidently closely related to *C. alpinus* (Chaudoir) also from Chimborazo ; but the author's description of the striae does not accord with *C. laevilateris,* and the claw joint of the tarsi is without cilia.

32. *C. piccolus,* Chaudoir, Ann. Soc. Ent. Fr., 1878, p. 299.

Originally taken in Colombia by Steinheil. Two examples taken by Mr. Whymper on Cayambe at 12-14,000 feet agree very well with Chaudoir's description ; but according to the author's synoptical table the fourth joint of the posterior tarsi ought to be without external lobe. Mr. Whymper's insect has a distinct narrow pointed lobe ; it is, however, difficult of detection, and may possibly have been overlooked. I am unwilling to consider it as a distinct species. An example sent me by M. Putzeys (who supplied M. de Chaudoir with specimens) as *C. piccolus,* is extremely like Mr. Whymper's, but the basal plica of the elytra is obsolete and the base so narrow that the humeral angles are brought near to the base of the thorax ; it cannot therefore be Chaudoir's species, and was evidently mistaken for it by Mr. Putzeys owing to its close resemblance.

33. *C. diopsis*, n. sp.

Hab. Pichincha (14-15,000 feet).

Intermediate between *C. piccolus* and *C. steno*, some examples almost bridging over the difference between the two. But the elytra instead of being rather broadly ovate as in *C. piccolus* are narrow oblong-ovate almost as in *C. steno*. It may be distinguished from *C. steno* by the elytral shoulders being somewhat distant from the angles of the thorax and the striæ being distinctly punctulated as in *C. piccolus*. Length, 8-8½ millim.

34. *C. steno*, n. sp.

Hab. Cayambe (15,000 feet), Pichincha (14-15,500 feet). Fourteen examples.

C. piccolo affinis sed multo angustior. Gracilis, castaneo-fuscus, palpis antennis pedibusque rufioribus: capite anguste ovato, post oculos tumido rotundato-angustato, collo parum depresso: thorace cordato-quadrato postice valde sinuato fere constricto angulis exstantibus acutis, margine laterali postice reflexo-elevato: elytris elongato-ovatis, plica basali integra parum arcuata, humeris a thorace distantibus apice obtuse rotundatis, punctulato-striatis, striis 5-7 subtilioribus: tarsis articulo 4to emarginato, posticis angulo exteriori elongato acuto, articulo 5to haud ciliato.

Long. 7-10 millim.

Var. *C. retentus*.

Hab. Cayambe (15,000 feet). Two examples.

Differs from the type only in the sides of the thorax continuing straight, or parallel from the sinuation to the hind angle, the latter remaining acute as in the typical examples. Long. 10 millim.

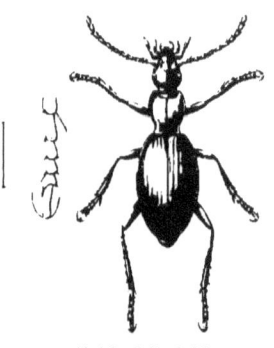

COLPODES STENO, BATES,
CAYAMBE & PICHINCHA.

35. *C. heberculus*, n. sp.

Hab. Between Antisanilla and Piñantura (11,000 feet). One example.

C. mosto (Dej.) quoad formam similis et affinis, sed minor colloque crasso. Paullo gracilior, elongato-ovatus, castaneo-fuscus, palpis antennisque rufioribus: capite subtriangulari, collo lato crasso, mox pone oculos transversim leviter depresso: thorace sat late subcordato-quadrato, ante basin subsinuato, angulis posticis rotundatis: elytris convexis, apice obtusis parum sinuatis, humeris subrectis, acute punctulato-striatis, interstitiis planis tertio impunctato: tarsis robustis, 4 posticis ad latera bisulcatis, articulo 4to intermediis bilobato, posticis emarginato angulo exteriori elongato.

Long. 8 millim. ♂.

Allied to the common *C. mœstus* of the Mexican plateaux, and belonging to the same section, but much smaller, a little narrower, and the head very different in the form of its broad convex neck, which has above a slight transverse impression: the eyes are convex, but the cheeks narrow rapidly behind them to the broad neck. The labrum has a straight fore margin, the palpi have very slightly fusiform apical joints, and the antennæ have the joints 5-11 notably broad and compressed. The claw joint, which, like the other joints of the tarsi, is shorter and stouter than usual in the genus, is naked beneath. The colour of the head and thorax above is nearly black, the elytra are chestnut-brown, and the body beneath as well as the legs, antennæ, etc., reddish; but the colour of the underside is probably variable as in other species of the section.

36. *C. Drusillus*, n. sp.

Hab. Pacific slopes (7-8000 feet). Two examples.

Gracilis, elongato-ovatus, nigro-piceus nitidus (♂ ♀) palpis antennis pedibusque rufioribus: capite sat gracile ovato post oculos gradatim angustato, collo supra parum depresso: thorace anguste ovato subquadrato, ante basin paullo sinuato, angulis posticis obtusissimis: elytris apice parum sinuatis, punctulato-striatis interstitiis planis 3to tripunctato: tarsis subtus dense tenuisetosis, 4 posticis bisulcatis articulo 4to bilobato lobo exteriori (præcipue posticis) elongato; 5to nudo.

Long. $8\frac{1}{4}$ millim. ♂ ♀.

Also allied to *C. mœstus*, but very much smaller and more slender, with rather narrow oval thorax having obtuse, nearly rounded, hind angles preceded by a slight sinuation of the sides and oblique basal margin near the angles. It thus resembles to deception certain slender species of *Agonum*. The elytra are elongate-ovate; the legs rather slender with deep sulci on the sides of the four hinder tarsi, the outer lobe of the fourth joint in the hindmost pair being narrow and elongated. The elytra have a distinct scutellar striole, and the first stria has at its origin an ocellated puncture.

37. *C. alticola*, n. sp.

Hab. Hacienda of Antisana (13,300 feet), Pichincha (12-13,000 feet), Antisanilla to Piñantura (11,000 feet), Machachi (9-10,000 feet). Numerous examples.

Elongata, antennis pedibusque sat gracilibus capite mox pone oculos citius angustato: niger politus, elytris ♀ opacis, antennis palpis (pedibusque interdum) rufioribus : thorace quadrato, postice angustato, lateribus antice plus minusve rotundatis ante basin leviter sinuatis, angulis posticis valde obtusis sed distinctis margine basali prope angulum obliquo ; elytris elongato-ovatis, acute subpunctulato-striatis, interstitiis planis tertio 3-5 punctato : tarsis 4 posticis extus sulcatis, subtus dense tenuisetosis, articulo 4to intermediis bilobato, lobo exteriori paullo longiori, posticis emarginato angulo exteriori distincte elongato ; articulo 5to subtus ciliato.

Long. 10½-12½ millim.

Agrees in some respects both with *C. alpinus* (Chaud.) from Chimborazo, and *C. dyschromus* (Chaud.) from Colombia, but the author's descriptions founded on too few individuals, and imperfect in other respects, leave one in doubt. I do not find an insect amongst Mr. Whymper's numerous series having simply emarginated fourth joint to the middle tarsi like *C. alpinus*, as Chaudoir's description implies, and the number of large punctures on the third interstice is not 3 (except in rare instances) as in *C. dyschromus* but 4 or 5 ; they are very variable in position, and are very conspicuous on the sericeous-opaque elytra of the ♀. The male is unusually brilliant in its black colour. The thorax varies greatly in width and rotundity of the sides.

38. *C. purpuratus*, Reiche, Rev. Zool., 1842, p. 375 ; Chaudoir, Ann. Soc. Ent. Fr., 1878, p. 340.

A common insect in Colombia. Mr. Whymper obtained one example from Milligalli (6200 feet).

This is the only species of *Colpodes* collected by Mr. Whymper belonging to the third and most numerous section of the genus, characterised by its long and narrow metathoracic episterna and almost universally metallic colours. All the other species have very short episterna, and black or dark-brown colours.

Subfam. TRECHINÆ.

39. *Trechus* ——— ? One broken specimen, undoubtedly of this genus, was taken on Cayambe at 15,000 feet.

Subfam. BEMBIDIINÆ.

40. *Bembidium fulvocinctum*, n. sp.

Hab. Tortorillas to Chuquipoquio, Chimborazo (12-13,000 feet) ; Hacienda of Antisana (13,300 feet) ; Valley of Collanes, Altar (12,500 feet) ; Pichincha, first camp (14,000 feet) ; Cayambe (15,000 feet). Seven examples.

Convexum apterum, nigro-piceum vel castaneum æneo-tinctum, elytris plus minusve distincte fulvo-marginatis, antennis basi pedibusque fulvo-rufis : sulcis frontalibus modice impressis sat vagis versus oculos curvatis : thorace rotundato-cordato versus basin valde angustato, angulis posticis acutis, fovea obliqua curta profunda utrinque ab angulo carina obtusa separata : elytris ovalibus versus humeros valde rotundatis ; striato-punctatis interstitiis planis tertio punctis parvis duobus ; striis versus apicem plus minusve obsoletis : stria 8va ante basin cum stria 9na conjuncta.

Long. 3½ millim.

In its very oval elytra resembles the St. Helena *Bembidia* more than any group of Europe or North America. The curve of the sides continues to the end of the plica which forms a little angle near the basal angle of the thorax. The general form is rather short and oval, the surface glossy, the elytral striae scarcely impressed, and the width of the tawny border very irregular.

41. *B. (Peryphus) Chimborazonum*, n. sp.

Hab. Tortorillas to Chuquipoquio, Chimborazo (12-13,000 feet). Eight examples.

B. decoro (Panz.) quoad formam haud dissimile sed apterum, elytris versus basin magis attenuatis, basi angustis, angulis humeralibus nullis ; olivaceo-nigrum scapo rufo, pedibus fulvo-piceis : oculis minus prominentibus colloque sat crasso : thorace cordato-quadrato, antice rotundato prope basin sinuato-angustato, angulis posticis acutis, fovea utrinque basali unica profunda ab angulo carina acuta separata : elytris auguste elongato-ovatis versus basin angustatis, striis 1-4 haud acute impressis sed interstitiis convexis, 5 indistincta 6-7 obsoletis, 8va fere usque ad basin extensa ; interstitio punctis magnis transversis duobus.

Long. 5-5½ millim. ♂ ♀.

Approaches in form the species allied to *B. decorum*, but the elytra narrowed in a moderate curve to the base without trace of humeral angles, and the membranous wings wanting. The frontal furrows are moderately long ; the punctures of the third elytral interstice in well-developed examples extend across the interstice ; the striae are scarcely at all impressed, and are rendered apparent only by the convexity of the interstices.

42. *B. (Notaphus) Cayambense*, n. sp.

Hab. Village of Cayambe (9320 feet). Two examples.

B. Aubei (Solier) affine sed magis depressum ; aeneum, immaculatum, scapo pedibus rufo-testaceis aeneo-tinctis : foveis frontalibus simplicibus latis ; thorace antice paullo magis quam postice angustato, angulis posticis subacutis, basi utrinque fovea lata, alteraque angusta ab angulo carina separata : elytris oblongis, punctato-striatis striis vix impressis, interstitiis planissimis 3^{io} et 5to caeteris latioribus, 3^{io} punctis duobus in medio interstitio sitis.

Long. 5 millim.

A distinct species of the very numerous and universally distributed group *Notaphus*, approaching nearest *B. Aubei* of Chili, though very different in colour and more flattened in form, in these respects having a general resemblance to the much larger Mexican species *B. placitum*.

Subfam. DRYPTINÆ.

43. *Galerita ruficollis*, Dejean, Sp. Gen. Col., i, p. 191.

Hab. Guayaquil. A widely-distributed species in Central America and the West Indies.

Tribe LAMELLICORNIA.

Fam. COPRIDÆ.

44. *Uroxys elongatus*, Harold, Col. Hefte, iii, p. 44.

Hab. Cotocachi (11-13,500 feet). Two examples. Previously known from Quito.

45. *U. latesulcatus*, n. sp.

Hab. Pichincha (12,000 feet), Machachi (9-10,000 feet). Six examples.

UROXYS LATESULCATUS, BATES
PICHINCHA AND MACHACHI.

U. elongato proxime affinis; differt elytris late sulcatis, sulcis opacis striis vix perspicue punctulatis interstitiis convexis politis: oblongus, niger, capite supra curvatim ruguloso, vertice unituberculato, clypeus bidentato; thorace transversim quadrato, angulis posticis rotundatis, sulco obliquo submarginali spatioque inter sulcum et marginem laterali convexo, fovea impresso; tibiis 4 posticis extus medio dente valido instructo.

♂ Pedibus anticis elongatis, tibiis valde curvatis, intus longe ante basin unidentatis.

Long. 12-15 millim. ♂ ♀.

16. *Ontherus aquatorius*, n. sp.

Hab. Ecuador, probably Pacific slopes.[1] One example only.

Elongato-oblongus, niger nitidus, antennis palpisque rufis; clypeo antice ruguloso, obtusissime bidentato, cornu verticis robusto, conico obtuso; thorace

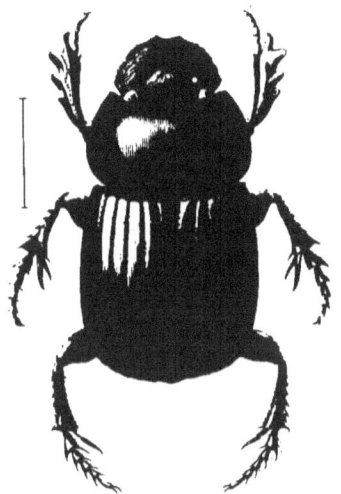

ONTHERUS ÆQUATORIUS, BATES.

relative brevi, antice fere verticaliter declive ibique et lateribus crebre punctulato, disco punctulis minus impressis, sulco dorsali vix impresso; elytris elongatis profunde striatis, striis sat grosse crenulatis, interstitiis elevatis sed supra late planulatis, subtiliter punctulatis; pectore et ventris lateribus grosse punctatis; metasterno lævi, medio fovea rotundata.

Long. 18 millim. ♂?

[1] The locality was lost by the setter.—*E. W.*

47. *Pinotus Satanas*, Harold, Col. Hefte, ii, p. 98.

Hab. Nanegal (3-4000 feet). Two examples. A common insect at moderate elevations in Colombia.

48. *P. Cotopaxi*, Guérin, Verhandl. Zoo. Bot. Verein, Wien, v, p. 588.

Hab. Pacific slopes (below 1400 feet). A single example. Appears to be peculiar to Ecuador.

49. *Phanæus conspicillatus*, Fabr., Syst. Eleuth., i, p. 32.

Hab. Nanegal (3-4000 feet). Numerous examples. Widely distributed in the northern part of South America, and a common insect in the forest plains of the Upper Amazons.

Fam. TROGIDÆ.

50. *Trox suberosus*, Fabr., Syst. Entom., p. 31.

Hab. Machachi (9-10,000 feet). A single example. Widely distributed throughout Tropical America.

51. *Cloeotus tubericauda*, n. sp.

Hab. Ecuador (altitude unknown [1]).

CLŒOTUS TUBERICAUDA, BATES.

C. metallico (Harold) valde affinis. Nigro-æneus politus corpore subtus rufo, clypei elytrorumque marginibus pedibusque (partim) rufescentibus: clypeo sat angulato genis acutis, fronte medio tumido sublævi cæteris grosse punctatis; thorace punctulato, lateribus late depressis grossius punctatis, disco postice lævi sulco antico submarginali integro, angulis posticis rotundatis; elytris (margine haud crenato) striato-punctulatis interstitiis planissimis, lævibus, 1mo apice 10mo 12mo longe ante apicem acute carinatis cæteris confuse et obtuse tuberosis; tibiis posticis apice extus productis tarsis elongatis.

Long. 6 millim.

Fam. MELOLONTHIDÆ.

52. *Astæna producta*, n. sp.

Hab. Machachi to Pedregal (10,000 feet). A single example.

Elongato-oblonga castaneo-rufa polita glabra, clypeo sat elongato, quadrato grossissime punctato, margine antico elevato, angulis fere rectis; thorace

[1] The locality was lost by the setter. — E. W.

antice valde angustato angulis longe productis acutis, sparsim grosse punctato: elytris punctulatis haud striatis interstitiis hic illie leviter convexis: pedibus gracilibus tarsis valde elongatis: pygidio opaco punctato, cum ventro pubescenti, medio pectore longius hirsuto.

♂ Antennis 8-articulatis, clava elongata gracile: tarsis anticis nec dilatatis subtus sparsim ciliatis.

Long. 14 millim. ♂.

Belongs apparently to Burmeister's Section II. A(2)a, but differs from all the species described by him in the greatly narrowed anterior part of the thorax and much produced anterior angles.

53. *Clavipalpus Antisanae*, n. sp.

Hab. Hacienda of Antisana (13,300 feet). Five examples.

Oblongus fulvo-castaneus, capite thorace scutello pygidioque nigronitidis hoc convexo glabro lævissimo: antennis 10-articulatis rufis, palpis maxillaribus articulo terminali elongato fusiforme, clypeo grosse sparsim

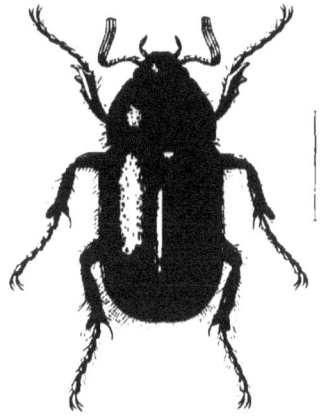

CLAVIPALPUS ANTISANAE, BATES.
HACIENDA OF ANTISANA.

alveolato-punctato rufescente, antice emarginato, vertice grosse confluenter punctato: thorace sparsissime punctato interdum macula marginali rufo: scutello laevi: elytris haud profunde punctulatis et transversim rugulosis, obsolete costatis: corpore subtus, pygidio excepto, pedibus, capite supra, thoracis elytrorumque marginibus fulvo-hirsutis.

Long. 17 millim. ♂.

Differs from *C. ursinus* by its 10-jointed antennae, slender fusiform terminal joint of the palpi, and polished, glabrous black pygidium, which has a few punctures only at the apex. The head is very slightly contracted just before the eyes and scarcely dilated in front of the contraction.

The following may perhaps be the ♀ of this species, but the single specimen being destitute of anterior tarsi its sex cannot be determined:—

♀ ?—*A C. Antisance* ♂ differt clypeo multo breviori mox ab oculis angustato, pygidio rufo passim sparse punctato.

Hab. Ecuador.

54. *C. Whymperi*, n. sp.

Hab. Machachi (9-10,000 feet), and between Machachi and Pedregal (10,000 feet). Six examples.

Paullo minor nitidus elytris rufo-castaneis, capite thoraceque nigris; capite omnino discrete punctato: thorace sparsissime punctato: scutello lævi: elytris parum profunde punctato-rugulosis: pygidio glabro, polito castaneofusco: corpore subtus antennis pedibusque fulvo-rufis et fulvo-pilosis: palpis maxillaribus articulo terminali sat gracili fusiforme; clypeo apice late rotundato sinuato: antennis 9-articulatis.

Long. 15-16 millim.

Var. ?:—clypeo parvo, semicirculari, antice haud perspicue sinuato: antennis 10-articulatis. ♂.

Differs from *C. Antisance* by its darker elytra—chestnut-reddish, instead of tawny-brown, and by their being more densely and distinctly punctulaterugose, though quite as glossy, and by the much less distinctly elevated costæ.

The examples with short rounded clypeus and 10-jointed antennæ are smaller and rather more ovate; there appears to be no difference in the anterior tibiæ or tarsi and very little in the abdomen. In colour and sculpture they are the same as the others.

Var. *C. Chimborazonus*. Differt a typo tantum antennarum clava fusconigra.

Hab. Tortorillas to Chuquipoquio, Chimborazo (12-13,000 feet). Two (♂) examples with 9-jointed antennæ.

55. *Ancylonycha austera*, Erichson, Wiegmans Archiv, 1847, i, p. 101.

Hab. Guayaquil. Two examples, taken indoors. Recorded by Erichson from Peru.

Fam. RUTELIDÆ.

56. *Anomala chloroptera*, Burm., Handbuch d. Ent., iv, 1, p. 262.

Hab. Guayaquil. One example, taken indoors. Previously known from Colombia (Burm.).

57. *Thyridium impunctatum*, n. sp.

Hab. Nanegal (3-4000 feet). One example.

Antichira chrysidi (Lin.) simile sed paullo magis ovatum, supra prasinum glabrum, subtus pedibusque subcupreo-auratum tarsis (viridi-auratis) pilis longis fulvis haud dense vestitis: clypeo thoraceque lateribus obsolete punctulatis, cætera superficie impunctata elytris striis nonnullis valde obsoletis; scutello sat parvo lateribus curvatis: pygidio irregulariter transversim aciculato.

Mandibulæ extus rotundatæ: processus sternalis elongatus, gracilis, descendens apice incrassatus; pedum 4 posteriorum ungues simplices.

Long. 22 millim. ♂.

Allied to *Th.* (*chlorota*) *euchloroïdes* (Murray), but smaller, more oblong in form, the sternal process longer and thinner, and the scutellum shorter and more shield-shaped, *i.e.* the sides regularly rounded. The hind margin of the thorax is not emarginated or sinuated on each side of the scutellum. It appears from the description to come very near the *Chlorota Bogotensis* of Kirsch, but differs (besides other points) in the punctuation of the head, the clypeus in Kirsch's species being "dense ruguloso-punctato," and in *Th. impunctatum* having only a few very faintly-impressed punctures distant from each other. I adopt Waterhouse's definition of the two allied genera, under which *Ch. Bogotensis* would rank as a *Thyridium*.

58. *Lasiocala fulvohirta*, Blanchard, Cat. de la Coll. d. Mus., p. 220.

Hab. Machachi (9-10,000 feet), Machachi to Pedregal (10,000 feet). Three examples.

59. *Platycœlia prasina*, Erichs., Wiegm. Archiv, 1847, i, p. 100.

Hab. Nanegal (3-4000 feet)—Peru (Erichson); Merida (Burmeister).

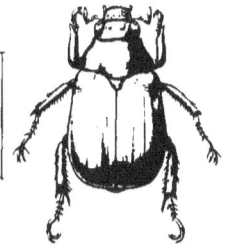

PLATYCŒLIA PRASINA, ERICHS.
NANEGAL.

60. *P. nigricauda*, n. sp.

Hab. Cotopaxi (12,000 feet). Four examples.

Elongato-oblonga, subparallelogrammica, viridescenti-ochracea, metasterno ventrisque segmento apicali fusco-nigris; capite sat angusto, clypeo creberrime punctato (punctis majoribus minoribusque intermixtis) fronte parce punctato, vertice lævi; thorace mox pone angulos anticos rotundato-dilatato, supra impunctato, nitido; elytris punctulato-striatis (interdum striis omnino obsoletis) interstitiis 3 et 5 elevatis; corpore subtus (præcipue pectore) pedibusque longe fulvo-hirsutis.

Processus sternalis brevis, acuminatus.

♂. Tibiæ anticæ 3 - ♀ 2-dentatæ.

Var. immatura? metasternum ventrisque segmentum terminale pallidius fusca.

Hab. Hacienda of Antisana (13,300 feet). One example.

LEUCOPELÆA, nov. gen.

G. *Platycœlia* affinis; differt corpore magis ovato elytrisque relative brevioribus (habitu G. *Arcadæ* similis), clypeo brevi, sutura arcuata; processo sternali nullo, mesosterno inter coxas angusto convexo; tarsis (unguibusque) sat gracilibus, articulis singulis longulis basi angustatis, ungue exteriori omnibus ante apicem denticulo parvo; corpore subtus ventroque minus planato.

61. *Leucopelæa albescens*, n. sp.

Hab. Cotopaxi (12,000 feet), Machachi (9-10,000 feet). Thirteen examples.

Testaceo-alba, corpore subtus fulvo-hirsuto; clypeo confluenter punctato; antennis rufis; thorace mox pone angulos anticos obtusissimos rotundato-dilatato, lævi; elytris striato-punctulatis, punctulis fuscis interdum obsoletis, interstitiis parum convexis. Long. 22-27 millim. ♂ ♀. For Figure see the Plate facing p. 31.

Fam. DYNASTIDÆ.
BAROTHEUS, nov. gen.

Gen. *Cyclocephala* (§ *Ancognatha*) proxime affinis; differt inter alia fronte utroque sexu tuberculo valido mediana armata. Corpus oblongum, robustum. Clypeus parabolicus antice acuminatus reflexus. Antennæ 10 articulatæ clava (♂) brevi. Labrum (sub clypeo obtectum) sat elongatum, corneum, medio carinatum. Maxillæ edentatæ, omnino pilosæ. Mandibulæ elongatæ productæ angustæ apice obtuse acuminatæ. Mentum apice angustatum acuminatum. Tibiæ anticæ breves et latæ, tridentatæ (♂ dentis obtusis). Tarsi antici (♂) articulo unguiculari incrassato, unguibus valde inæqualibus, grossiori apice bifido.

The stout frontal tubercle, or short horn, is formed by the thickening and folding of the suture which separates the clypeus from the forehead, and it is

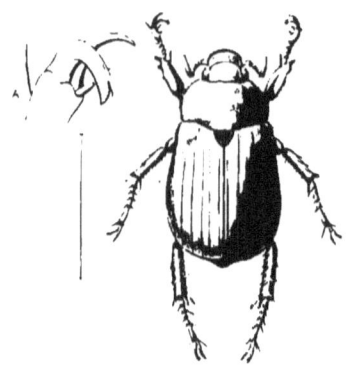

A. ANTERIOR CLAW OF THE MALE.
FROM NORTH-WEST SIDE OF COTOPAXI. 12 000 FEET

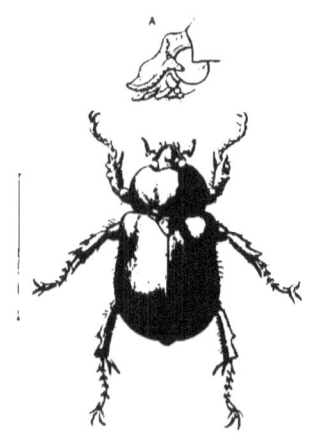

A SIDE VIEW OF HEAD.
FROM MACHACHI, CHIMBORAZO, ETC. 9000-11,700 FEET

perpendicular, or even concave on its anterior face and gradually sloping behind, the frontal suture being traceable to its summit. It is peculiar in being equally developed in both sexes.

62. *Barothens Andinus*, n. sp.

Hab. Chillo (9000 feet); Machachi (9-10,000 feet); between Machachi and Pedregal (10,000 feet); Chuquipoquio, Chimborazo (11,700 feet). Eight examples.

Fusco-niger nitidus (♂ elytris sericeo-opacis): capite et thorace sparsim, clypeus crebrius, punctatis, elytris impunctatis interdum obsolete striato-punctatis: antennis, palpis tarsisque plus minusve rufis; interdum thorace et pedibus castaneo-rufis: pectore longe fulvo-hirsuto: pygidio obsolete punctulato et strigoso apice fulvo ciliato elytrorum marginibus breviter ciliatis. Long. 9-11 lin. ♂ ♀. For Figure see the accompanying Plate.

63. *Cyclocephala diluta*, Erichson, Wiegm. Archiv, 1847, i, p. 97.

Hab. Guayaquil; Pacific slopes (below 1400 feet). Two examples.

Erichson described the species as being without spots on the elytra. Mr. Whymper's examples have several ill-defined brownish streaks, in one more numerous than in the other, showing the variability of the character.

64. *C. collaris*, Burm., Handb., v, p. 47.

Hab. Pacific slopes (below 1400 feet). A single female example, closely resembling others with which I have compared it from Panama.

65. *C. rubescens*, n. sp.

Hab. Nanegal (3-4000 feet). Four examples.

C. lucidæ (Burm.) proxime affinis, sed differt inter alia clypeo haud sinuato. Elongato-oblonga convexa, testaceo-rufa glabra nitida, vertice thoracisque maculis quatuor (transversim seriatis) nigris: clypeo semiovato marginibus reflexis, vermiculato-rugulosis et ocellato-punctatis, fronte bituberculata, vertice sparsim punctato: thorace antice rotundato-dilatato lævissimo: elytris obsolete striato-punctatis, ♀ callo elongato submarginali margineque proximo incrassato: abdominis segmento penultimo dorsali elongato (♂ ♀) exserto, pygidio relative curto, convexo, nitido setifero-punctato. Long. 22 millim. ♂ ♀.

Belongs to Burmeister's section "*Cyclocephala parabolica.*" The elytra or rather the "after-body" is parallel-sided and not wider than the thorax.

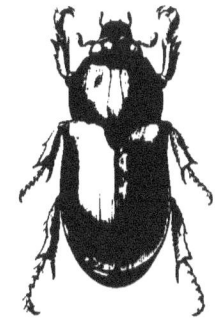

CYCLOCEPHALA RUBESCENS, BATES.
NANEGAL.

66. *Dyscinetus dubius*, Oliv., Ent., i, 5, p. 32, t. 3, f. 20, *a*, *b*; *geminatus*, Fab., Syst. El., ii, 166; Burm., Handb., v, p. 78.

Hab. La Mona (under 200 feet). Four examples. Widely distributed in Equatorial America, throughout the Amazons plains, Cayenne, etc.

Burmeister drew up his description of *D. geminatus* evidently from South Brazilian specimens, in which the pygidium is coarsely shagreened for a greater or less extent at the base in both sexes and sparsely punctured in the rest of its extent. In the true Guiana form of the species, to which Olivier's and probably Fabricius' specimens belonged, the pygidium in the male is densely rugulose or vermiculate-punctate throughout, except for a small space on the disk, and in the female sparsely punctured with the base roughened. The S. Brazilian form is more oblong (the elytra being scarcely dilated behind), the ♂ anterior claws and claw joint much thicker, and the clypeus generally more coarsely punctured than in the Guiana type; in the latter the ♂ pygidium is clothed with stiff hairs, the ♀ naked and much more compressed laterally.

Ecuadorian examples are much more finely punctured than those from the Amazons.

67. *Stenocrates laborator*, Fabr., Syst. Ent., i, p. 18.

Hab. Guayaquil. Three examples. Also Amazons plains and Surinam.

68. *S. holomelanus*, Germar, Ins. Spec. nov., p. 116.

Hab. Pacific slopes (below 1400 feet). Three examples. Found throughout tropical South America.

69. *Megaceras Philoctetes*, Oliv., Ent., i, 3, p. 16, t. 14, f. 125 : syn. *M. Toucer*, Burm., Handb., v, p. 223.

Hab. Nanegal (3-4000 feet). One female example, probably of this species.

BARYXENUS, nov. gen.

Ad Sect. *Pimelopides* pertinet. Corpus oblongum, crassum, convexum. Caput parvum; clypeus triangularis; frons (♀) tuberculo acuto armata. Mandibulæ breves, crassæ, obtusæ extus rotundatæ. Maxillæ subobtusæ, inermes, pilosæ. Mentum antice angustatum. Palpi articulo terminali cæteris conjunctis longiori subcylindrico. Antennæ articulis 3-5 brevibus 6-7 latæ. Thorax antice parum retusus tuberculis duobus antico-discoidalibus. Tibiæ anticæ 4-dentatæ; intermediæ et posticæ intus planatæ, extus unicarinatæ apice oblique truncatæ, truncaturæ marginibus simplicibus; calcaria lata, obtusa: tarsi articulo 1mo obtuso triangulari. Prosternum lobo postcoxali munitum.

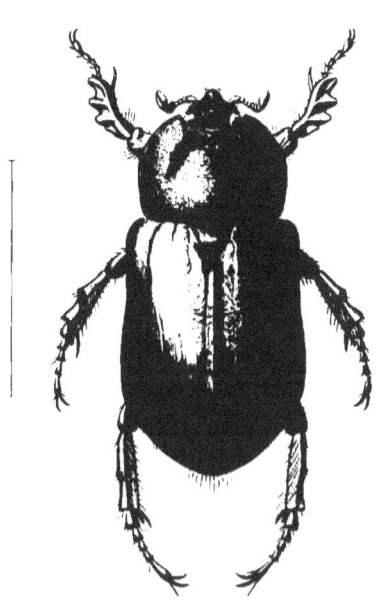

BARYXENUS ÆQUATORUS, BATES.
FROM MACHACHI, 10,000 FEET.

LIBRARY OF
U. S. GRANT IV

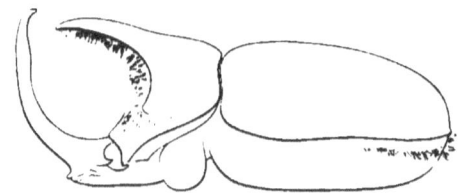

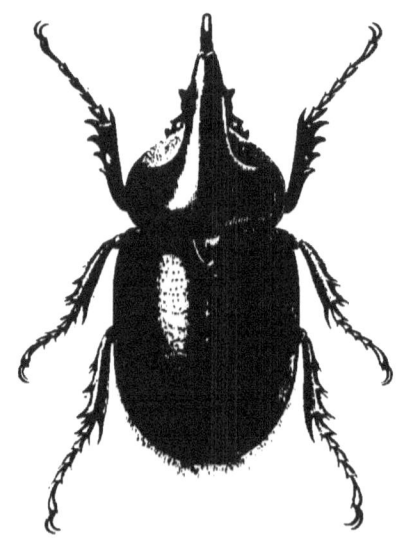

FROM THE COUNTRY TO THE WEST OF QUITO

70. *Baryxenus æquatorius*, n. sp.

Hab. Machachi (9-10,000 feet).

Castaneo-fuscus, capite et thorace antice arcuatim-strigulosis: elytris nitidis vage rugatis et hic illic obsolete striato-punctatis, stria suturali paullo distinctiori: pectore rufo-hirto: pygidio politissimo, basi punctulato. Long. 40 millim. ♀.

A singular form, of which unfortunately only a single female specimen was obtained. Although belonging to Lacordaire's group *Pimelopides*, the hind tibiæ are not so broad and robust as in the other genera of that group, and the edge of the obliquely truncated tibiæ is somewhat flexuous. The propygidium is not produced (at least in the ♀), and the pygidium is very large and convex. For Figure see the Plate facing p. 32.

71. *Heterogomphus Bourcieri*, Guérin, Rev. Zool., 1851, p. 160.

Hab. Between Guallabamba and Guachala (9000 feet).

72. *H. Whymperi*, n. sp.

H. Schönherri (Burm.) affinis: differt statura majori elytris basi lævibus, ♂ cornu thoracico vix ascendenti, valde elongato, versus apicem angustato apice breviter emarginato.

Long. 30 lines = 64 millim. ♂.

Hab. West of Quito (height not known).

Mr. Whymper obtained at Quito one specimen only of this fine species, which differs too much from the Colombian *H. Schönherri* to be considered merely an extreme development of that insect. The general form is longer and less convex, the elytra relatively longer, and the thorax more gradually narrowed in front. It is glossy-black above and dark rusty-black beneath. The head horn is much longer than in *H. Schönherri* ♂, less regularly curved, flexuous, and thickened behind a little before the apex; the thoracic horn is less elevated, more horizontal and greatly lengthened, the apex reaching beyond the clypeus; it is densely clothed, as well as the punctured anterior concavity of the thorax, with reddish hairs, which also cover a great part of the under surface of the body and the base of the pygidium. The sides of the thorax are coarsely rugose-punctate, the upper surface smooth. The elytra are nearly smooth near the scutellum and densely vermiculate-rugose and punctate, but much less coarsely than *H. Schönherri* in the rest of their surface. For Figure see the accompanying Plate.

73. *Enema Pan*, Fabr., Syst. Ent., i, p. 5; Burm., Handb., v, p. 235.

Hab. Nanegal (3-4000 feet). One male and one female example.

74. *Strategus Aloeus*, Lin., Syst. Nat., i, 2, p. 542 ; Burm., Handb., v, p. 235.
   Hab. Nanegal (3-4000 feet). One example.

75. *Golofa Ægeon*, Fabr., Syst. El., i, p. 4 ; Burm., Handb., v, p. 253.
   Hab. Milligalli (6200 feet). One pair.

PRAOGOLOFA, nov. gen.

A Gen. *Golofa* differt capite thoraceque ♂ inermibus. Quoad formam *Golofis* ♀ similis. Clypeus antice attenuatus, apice brevissime bidentatus : frons medio ♂ tuberculo conico, ♀ carinula transversa : mandibulae extus valde sinuatae apice latae, obtuse bidentatae, dente interiori multo minori. Thorax ♂ et ♀ omnino inermis : pedes et ungiculi (praecipue ♂) sat elongati, tibiae anticae ♂ tridentatae, ♀ quadridentatae (dente 4to parvo); tibiae posticae margine apicali leviter flexuoso sed haud dentato, breviter setoso.

Notwithstanding the absence of the remarkable horn-like projections from the head and thorax so characteristic of the genus *Golofa*, and which have been considered as essentially distinguishing the sub-family to which that genus belongs, there can be no doubt of the correctness of referring the present form to the same group. The form and sculpture of the body, the parts of the mouth and other characters are as in *Golofa*, and the legs and tarsi differ in relative length only in degree. A species of *Golofa* with unarmed thorax, apparently undescribed, connects the genus with the typical *Golofa*. The species here described somewhat resembles the *Cyclocephalæ*.

76. *Praogolofa unicolor*, n. sp.
   Hab. Between Guallabamba and Guachala (9000 feet). Two examples.

Fulvo-testacea nitida, pectore, pygidio basi femoribus tibiisque longe et dense fulvo-pilosis : capite grosse et crebre punctato nigro-fusco : thorace sparsim punctulato : elytris paullo grossius substriato-punctulatis et rugosulis, costis indistinctis.
   Long. 33-35 millim. ♂ ♀.

Golofa inermis, Thomson, is described as having the thorax and elytra narrowly bordered with black and the scutellum and suture black : the borders and scutellum in the present species are pale-coloured like the rest of the surface. For Figure see the accompanying Plate.

Fam. CETONIIDÆ.

77. *Gymnetis flavocincta*, n. sp.
   Hab. Nanegal (3-4000 feet). One example only.

G. Chevrolatii affinis ; differt elytrorum margine flavo rectissimo, ramulis

LIBRARY OF
U. S. GRANT IV

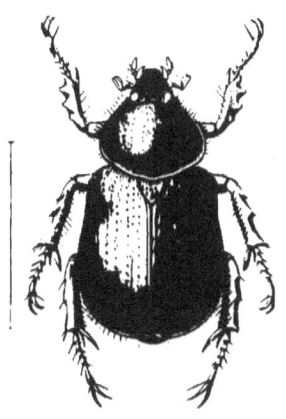

PRIOCOLOFA UNICOLOR, BATES.
FROM BETWEEN GUALLABAMBA AND GUACHALA, 9000 FEET.

nullis; Fuliginoso-nigra subtus nigra nitida; clypeo sicut in *G. Cherrolatii*
paullo ampliato, alte reflexo-marginato antice recto;
processo sternali antice minus acute tuberculato;
abdomine subtus fere lævi, lateribus punctis per-
paucis.

Long. 26 millim. ♀ ?

Bears some resemblance to *G. magnifica*, but
having the broader and more highly margined clypeus of *G. Cherrolatii* ♀. From both it differs in
the yellow border being very sharply and straightly
limited (without trace of branchlet) from the
shoulder to the suture, a short indistinct projec-
tion appearing only just beneath the apical callus.

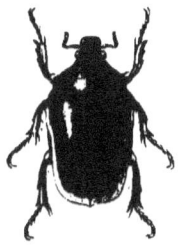

GYMNETIS FLAVOCINCTA, BATES.
NANEGAL.

Fam. PASSALIDÆ.

78. *Passalus furcilabris*, Eschsch., Nouv. Mem. Mosc., i, 1829, p. 25; Kaup, Col. Hefte, iv, p. 25.

Hab. Nanegal (3-4000 feet). One example. Also Cayenne.

79. *P. platyrhinus*, Hope, Cat. Lucanidæ, p. 28; Kaup, Col. Hefte, iv, p. 28.

Hab. Nanegal (3-4000 feet). Two examples. Also Colombia and Venezuela.

80. *Passalus* ——?

One example of a species of this genus from the Pacific slopes (5000 feet), which cannot satisfactorily be determined.

81. *Phoroneus punctatostriatus*, Percheron, Monogr. Passal., p. 78, t. 6, fig. 1.

Hab. Nanegal (3-4000 feet). One example. A common Central American species.

82. *P. binominatus*, Percheron, Monogr. Passal., Suppl., i, p. 23.

Hab. Nanegal (3-4000 feet). Two examples. Recorded from the West Indies.

83. *Phoroneus* ——?

A single specimen taken at Tanti (1890 feet [1]); the species indetermin-
able.

[1] On the authority of Father Menten. See Proc. Royal Geog. Soc., p. 189, Aug. 1881.—*E. W.*

84. *P. Maillei*, Percheron, Monogr. Passal., Suppl., i, p. 31, t. 78, fig. 6.

Hab. Pacific slopes (5000 feet). A widely-distributed species in the northern part of South America, and in Central America.

85. *Pertinax morio*, Percheron, Monogr. Passal., p. 83, t. 6, fig. 4.

Hab. Nanegal (3-4000 feet). Five examples. Recorded also from Colombia and Mexico.

86. *Ndeus punctatissimus*, Eschsch., Nouv. Mém. Mosc., 1829, i, p. 19.

Hab. Pacific slopes (below 1400 feet). One example. A common insect in Peru and Colombia.

Tribe LONGICORNIA.

Fam. *PRIONIDÆ*.

87. *Parandra glabra*, De Geer, Mém., iv, p. 351.

Hab. Chillo (9000 feet). One example. A widely-distributed species in Tropical America.

88. *P. luciana*, Thoms., Mus. Scientif., p. 86.

Hab. Milligalli (6200 feet); Nanegal (3-4000 feet). Four examples.

Three males and one female of a species which answers fairly well to Thomson's description of *P. luciana*.

89. *Prionocalus Whymperi*, n. sp.

Hab. Milligalli (6200 feet). Two ♂ and one ♀ examples.

♂. *P. Buckleyi* proxime affinis, differt tantum elytris versus apicem magis angustatis abdominis apicem tegentibus, sculpturaque toti corporis paullo subtiliori, creberrime et usque ad elytrorum apicem aequaliter vermiculato-rugosis; pedibus et palpis nigris.

Long. 40-45 millim.

♀. A *P. Buckleyi* ♀ valde differt; elytris valde elongatis, abdominis apicem longe transeuntibus, lateribus fere aequaliter rotundatis, apice suturali rotundato; sculptura sicut in ♂ crebriori et subtiliori quam in *P. Buckleyi*: pedibus et palpis piceo-nigris.

Long. 65 millim.

If the ♀ example above described really belongs to the ♂, this species is certainly distinct from *P. Buckleyi*, the elytra in *P. Buckleyi* ♀ being very much shorter and not covering the pygidium. The only characters dis-

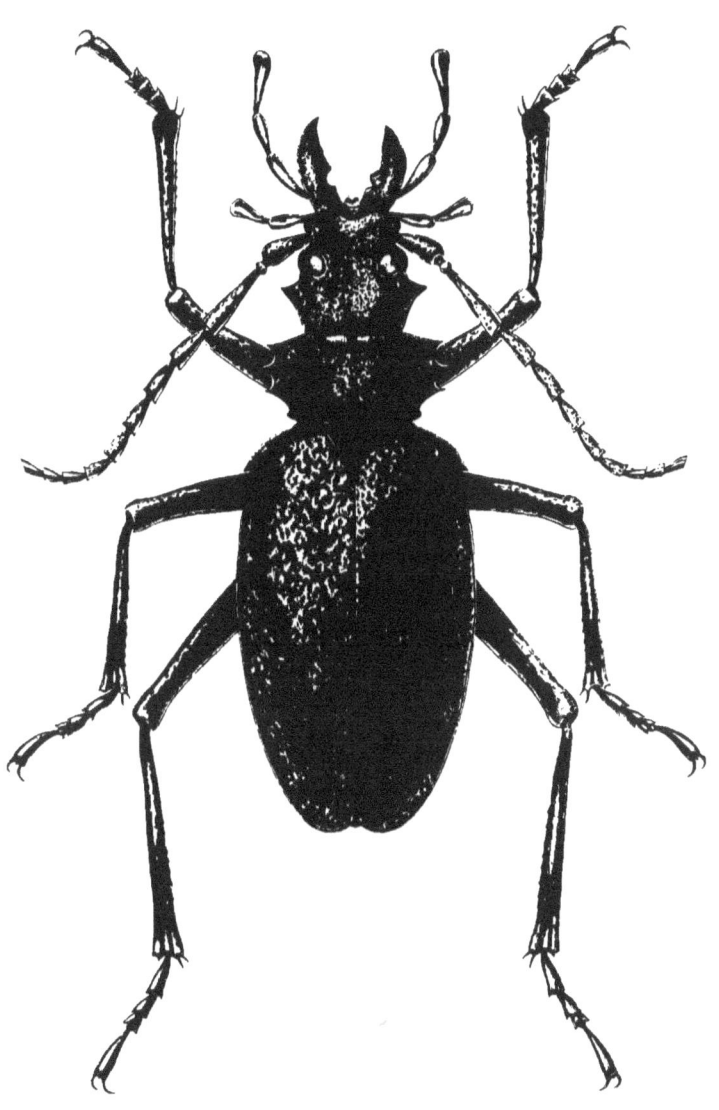

FROM MILLIGALLI. 6200 FEET.

tinguishing the males of the two forms are the flatter, less uneven, surface of the thorax, the more tapering elytra (almost as tapering as those of *P. atys* ♂), and the minuter, sharper, and more vermiculate rugosity of the whole surface, which on account of the sculpture is less glossy than in *P. Buckleyi*. For Figure see the accompanying Plate, which is on a scale one half larger than nature.

90. *P. Buckleyi*, Waterhouse, Entom. Monthly Mag., viii, p. 261 (Ap. 1872).

Hab. Pacific slopes (4000 feet). One ♂ example, which in the sculpture of the elytra is a little less coarse than in the typical form, and to that extent approaches *P. Whymperi*. In outline and in the shortness of the elytra it quite agrees with *P. Buckleyi*.

91. *P. trigonodes*, n. sp.

Hab. La Mona (under 200 feet), one ♂. Taken also by Mr. Buckley on his last journey.

♂ *P. Whymperi* affinis, sed angustior, elytris postice magis attenuatis, elongato-triangularibus, abdominis apicem haud transeuntibus; supra passim æqualiter creberrime vermiculato-rugulosus: tarsis omnibus, tibiis anticis palpisque castaneo-rufis. Variat antennis (ab articulo 2^{ndo}) castaneo-rufis, pedibusque interdum rufo-piceis. Long. 27-37 millim. ♂.

92. *Mallodon maxillosum*, Fabr., Syst. Ent., p. 163.

Hab. Guayaquil (taken indoors). One example, ♀.

Fam. CERAMBYCIDÆ.

93. *Chlorida cincta*, Guérin, Rev. Zool., 1844, p. 259.

Hab. Near Chillo (9000 feet). One example. Widely distributed in Central America, up to Mexico.

94. *Achryson lineolatum*, Erichson, Wiegm. Archiv, 1847, i, p. 142.

Hab. Guayaquil. One example (taken indoors). Inhabits also the coast-region of Northern Peru.

95. *Eburia quadrinotata*, Latr., Voy. de Humboldt & Bonpland, Zool., i, p. 165, t. 16, f. 9.

Hab. Guayaquil (taken indoors). A common insect in the north-western region of South America.

96. *Eurysthea angusticollis*, n. sp.

Hab. Machachi (9-10,000 feet), and between Machachi and Pedregal (10,000 feet). Three examples.

E. Lacordairei (Lac., Gen. Atlas, t. 87, fig. 1) affinis, sed multo angustior. Angusta, sublinearis, griseo-hirta, castaneo-fusca (elytris æneo-tinctis) antennis, pedibus maculisque (vel vittis) elytrorum testaceo-fulvis: thorace angusto fere cylindrico, grosse punctato, ochraceo-vario, tuberculis quinque parvis (mediano

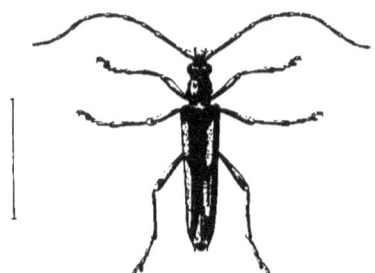

EURYSTHEA ANGUSTICOLLIS, BATES.
MACHACHI TO PEDREGAL.

et 2 posterioribus parvis vel obsoletis) spina laterali longa et acuta; scutello ochraceo-pubescenti: elytris crebre punctulatis, pilis incumbentibus alterisque erectis vestitis, utrinque vitta irregulari fulva apud medium et ante apicem furcata, apico utrinque sinuatis, angulo suturali producto, exteriori longe spinosa.

Variat elytris vitta medio interrupta.

Long. 17 millim.

The narrow thorax and its less elevated tubercles and sharp lateral spine distinguish the species from *Eu. obliqua* and *Eu. Lacordairei*.

97. *Cyllene elongata*, Chevrolat, Ann. Soc. Ent. Fr., 1861, p. 379.

Hab. Pacific slopes (1-3000 feet). One example.

98. *Trachyderes vermiculatus*, n. sp.

Hab. Chillo (9000 feet). Two examples.

Elongato-oblongus, supra planulatus, subtus dense griseo-hirtus, supra thorace dense elytris parcius pubescentibus; fuscescente-rufus, subopacus, antennis rufis 3-6 apice, 7-9 totis, nigris; pedibus rufis, tarsis nigris, vel totis nigro-fuscis; elytris sutura et apice interdum nigro-fuscis: thorace medio carina transversa, lateribus spina mediana elongata retrorsum curvata, tuberibusque duobus obtusis ante spinam; elytris vermiculato-rugosis; prosterno processu verticali anteriori parvo obtuso.

Long. 23-26 millim. ♀.

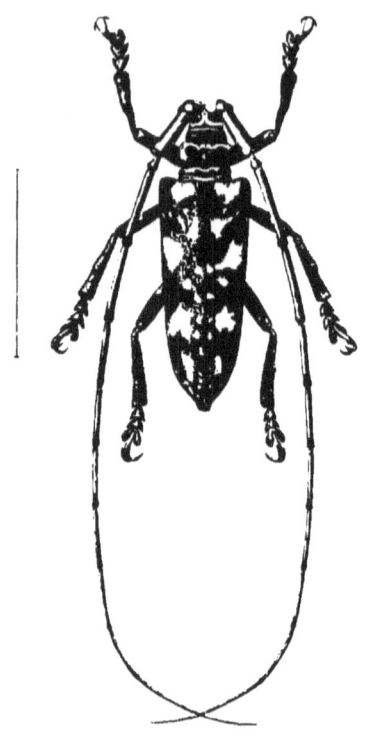

NEAR CHILLO, 9000 FEET

Nearest allied to the Brazilian *T. jureneus*, from which it differs greatly in markings. It is only in well-preserved examples that the elytra are seen to be pubescent, clothed with short laid hairs and long erect hairs, the latter most thickly near the base. See Figure on p. 6.

Fam. LAMIIDÆ.

99. *Tæniotes marmoratus*, Thomson, System. Cerambyc., p. 554.

Hab. Milligalli (6200 feet). Three examples.

100. *Hammoderus sticticus*, Bates, Trans. Ent. Soc. Lond., 1874, p. 225.

Hab. Near Chillo (9000 feet). One example. Taken also by Buckley, on the Morona, Ecuador. For Figure see the accompanying Plate.

101. *Oncideres callidryas*, Bates, Ann. Mag. N. H., xvi, 1865, p. 175.

Hab. Tanti (1890 feet). One example. A variety with less distinct and duller maculation of the elytra.

102. *Hypsioma* ——?

Hab. Nanegal (3-4000 feet). A single example, possibly of an undescribed species.

103. *Carneades nodicornis*, Bates, Entom. M. M., xvii, p. 277.

Hab. Milligalli (6200 feet). One example. Taken previously by Mr. Buckley in some numbers in Ecuador, and by Salmon in the Cauca Valley, Colombia.

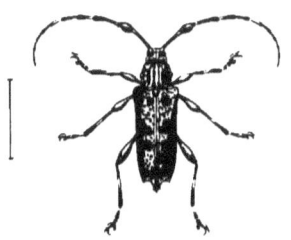

CARNEADES NODICORNIS, BATES.
MILLIGALLI.

COLEOPTERA—(CONTINUED).

By DAVID SHARP, M.B.

Fam. *DYTISCIDÆ*.

1. *Rhantus vicinus*, var. *Colymbetes vicinus*, Aubé, Sp. Gen., p. 243.

   Hab. Ecuador (locality unknown).[1] Previously found in Colombia.

   The single female differs slightly from the Colombian individuals in the sculpture of the thorax and the length of the hind claws, and when the male is known may possibly prove to be a distinct species.

Fam. *SILPHIDÆ*.

2. *Silpha Cayennensis*, Cast. Hist. Nat., ii, p. 5.

   Hab. Forests above the Bridge of Chimbo (1-3000 feet). Two examples. This is an abundant species in Equatorial America.

3. *S. microps*, n. sp.

   Hab. Quito (9500 feet). Six examples.

   Nigra; antennis brevibus, articulo octavo valde transverso ; prothorace dense punctato, medio lineis duabus posterius convergentibus elevatis ante basin desinentibus, basi utrinque linea brevi ; elytris crebre, fere fortiter, punctatis, lineis tribus elevatis argutis, externa pone medium abbreviata a callositate transversa limitata, apicibus nullo modo abbreviatis, angulo suturali anguste subrotundato ; abdominis segmento ultimo testaceo ; oculis minoribus. Long. corp. 16 mm.

   Two females were found of this species, which notwithstanding its very ordinary appearance does not appear to be closely allied to any other.

Fam. *STAPHYLINIDÆ*.

4. *Philonthus Whymperi*, n. sp.

   Hab. Hacienda of Guachala (9217 feet), Antisanilla to Piñantura (11,000 feet), Machachi (9-10,000 feet). Five examples.

   Niger, nitidus, capite thoraceque nitidissimis, subviolaceo-tinctis, elytris purpureis ; antennis sat elongatis, extrorsum haud crassioribus, articulo penultimo subquadrato ; capite quadrato ; thorace lateribus subsinuatis, posterius angustiore ; elytris prothorace longioribus et latioribus, crebre sat

---

[1] The catalogue number was lost by the setter.—*E. W.*

fortiter punctatis; abdomine basi parce, apice crebre, punctato; tarsis anterioribus utroque sexu simplicibus: maris segmento ultimo ventrali apice tantum leviter emarginato. Long. corp. 10-12 mm.

The five examples before me of this species are remarkable from the variation in the number of the thoracic punctures. These may be either 3, 4, or 5 in each series; thus it is somewhat uncertain to which of Erichson's groups of species it should be referred, but probably it may prove to belong to group 4. Mr. Whymper informs me that this species is rather abundant in the interior of Ecuador.

5. *P. divisus*, n. sp. (Group 5, Erichson).

Hab. Machachi (9-10,000 feet). One example.

Niger, nitidus, elytris læte flavis, basi violaceo; antennis parum elongatis, articulis quinto ad decimum transversis; capite oblongo, parce fortiter punctato; thorace nitidissimo lateribus subrectis, angulis anterioribus subrotundatis, utrinque serie discoidali punctorum quinque impresso; elytris parum profunde, sed haud subtiliter punctatis; abdomine nitidulo, parce punctato; tarsis rufescentibus. Mas incognitus; feminæ tarsis anterioribus simplicibus. Long. corp. circiter 12 mm.

6. *Sterculia impressipennis*, n. sp.

Hab. Nanegal (3-4000 feet). Three examples.

Cyaneo-violaceus; antennis sat elongatis, scapo gracili; capite latiore, densissime fortiter punctato; prothorace angusto, posterius parum latiore, crebrius fortiter punctato, posterius in medio lævigato, ad latera parum impresso; elytris crebre obsolete punctatis, versus latera anterius leviter, ad angulos posteriores profunde, impressis; capite subtus fortiter sat crebre punctato. Long. corp. 21 mm.

The three examples vary somewhat in colour and size, and a good deal in the length of the mandibles, but agree in the thoracic structure and punctuation.

7. *Cryptobium*, sp. ?

Hab. Pacific slopes (7-8000 feet). A single female example in a mutilated condition, unsuitable for description.

8. *Pæderus ornaticornis*, n. sp.

Hab. Guayaquil. A single example, taken indoors.

Niger, antennarum articulis tribus basalibus articuloque ultimo flavis, elytris viridis; antennis tenuibus sat elongatis; capite parum lato, sat crebre haud fortiter punctato; thorace oblongo-ovali, lateribus obsolete punctatis;

elytris parallelis, prothorace longioribus, crebre, fortiter subruguloso-punctatis, setulis nigris erectis adspersis; abdomine parce punctato. Maris segmento ultimo ventrali profunde exciso, excisione angulis fere rectis. Long. corp. 10 mm. Ex affinitate *P. cyanipennis* Guer. sed duplo minor, staturaque graciliore.

Fam. TENEBRIONIDÆ.

9. *Epitragus dilutus*, n. sp.

Hab. Bodegas (level of sea). Three examples.

Rufo-aeneus, nitidus, pube brevissima parcius adspersus, clypeo apice bisinuato, lobo intermedio late rotundato; prothorace crebre punctato, basi in medio vix perspicue impresso; elytris haud striatis, subtiliter seriatim punctatis, interstitiis obsolete punctulatis; corpore subtus infuscato; antennis pedibusque rufis. Long. corp. 8 mm.

The Sandwich island *E. diremptus*, Karsch, is very closely allied to this species.

10. *Nyctobates gigas*, (*Tenebrio gigas*) Linn. Syst. Nat., ed. 12. p. 674.

Hab. La Mona (200 feet). Two examples.

This species is very abundant in Tropical South America.

11. *Strongylium denticolle*, n. sp.

Hab. Milligalli (6000 feet). A single example.

Oblongo-ovatum, metallico-viride, antennis extrorsum nigris, femoribus basi flavo; prothorace transverso, inaequali, sat crebre inaequaliter punctato, lateribus in medio denticulatis, nullo modo marginatis; elytris posterius subacuminatis, ad basin et in medio seriatim grosse fossulatis, ad apicem profunde striatis; pedibus gracilibus, tarsis tenuibus elongatis, articulo ultimo praesertim elongato, quam articulo basali duplo longiore; corpore subtus laete metallico, versicolore, pectoris lateribus fulgidis. Long. corp. 16 mm.

This is a very aberrant species, but in the present unsatisfactory condition of the group of genera forming the *Strongyliides*, had better be treated as a member of the great genus *Strongylium*: the general form, however, approaches to *Spheniscus*: the metathoracic episterna are broader in front than in other species of the genus, but yet are more distinctly separated from the middle acetabula; the antennae are not elongate, the penultimate joint being scarcely longer than broad, and have quite the form found in some species of *Strongylium*: the clypeus is very definitely marked off. For Figure see the accompanying Plate.

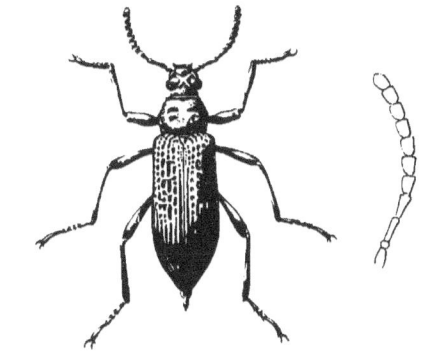

STRONGYLIUM DENTICOLLE, SHARP.
FROM MILLIGALLI, 6000 FEET.

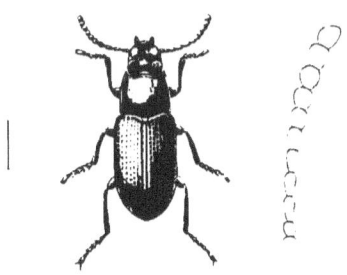

ASOPUS OPATROIDES, SHARP
FROM THE PACIFIC SLOPES, BELOW 1400 FEET.

ASIOPUS, nov. gen.

Mentum angustum, palpi labiales breves, articulo ultimo crassiusculo, ovali; palpi maxillares articulo ultimo magno, securiformi. Antennæ elongatæ, apicem versus parum incrassatæ, articulis ultimis subglobosis. Clypeus antice emarginatus.

This little insect has probably hitherto escaped the notice of entomologists on account of its obscurity. It has the size and appearance of an *Alphitobius* or *Opatrum*, but its systematic place must be at present in the group *Adeliides* of Lacordaire, near *Adelium*, from which it differs by the emarginate clypeus, and very narrow mentum. The tarsi are densely clothed with fine pubescence beneath; the anterior feet are rather broad and large, with the fourth joint excavate above for the reception of the terminal joint, but not bilobed. The prothorax is strongly margined at the sides, and the body is provided with wings. Probably the nearest true ally is the genus *Sciophagus*, which is itself of doubtful position.

12. *Asiopus opatroides*, n. sp.

Hab. Pacific slopes (below 1400 feet). Two examples.

Oblongus, parum convexus, subopacus, nigro-piceus, antennis pedibusque piceis; prothorace crebre punctato, basi utrinque sinuato, angulis posterioribus retrorsum spectantibus, acutis; elytris fortiter, regulariter seriatim punctatis, interstitiis subtilissime punctulatis. Long. corp. $7\frac{1}{2}$ mm.

The antennæ are as long as the head and thorax, with the third joint three times as long as the very short second joint, the seventh, eighth, and ninth joints are each about as long as broad, the tenth a little shorter. The body is destitute of pubescence, but the antennæ, tibiæ, and the punctures of the head bear excessively minute short grey specks of hairs or scales sufficient to give them a feebly grisescent appearance. The punctures in the series near the suture of the wing-cases are connected by fine longitudinal striæ, but this striate appearance is absent from the other series. For Figure see the Plate facing page 42.

13. *Meloe sexguttatus*, n. sp. (subgen. *Pseudomeloe*, F. & G.).

Hab. Pichincha (12,000 feet), Machachi (9-10,000 feet), Cotocachi (11-13,500 feet). Seven examples.

Niger, capite thorace elytrisque nitidis, per-parum punctatis, his maculis quatuor basalibus flavis, singuloque ad apicem macula majore aurantiaca, abdominis lateribus late testaceis; prothorace anterius vage transversim impresso, basi in medio longitudinaliter impresso, lateribus simplicibus, lævigatis. Long. corp. (insecto siccato) 14 mm.

The flavescent colour of the hind-body is confined to the membranous border in which the stigmata are situated, the dorsal and ventral plates being entirely black. The Peruvian *M. pustulatus*, Er., is an insect marked in a manner similar to *M. sexguttatus*, but has the thorax tuberculate and rugose at the sides.

14. *Amauca debilis*, n. sp.

Hab. Machachi (9-10,000 feet), Illiniza (14,000 feet). Two examples.

Sordide testacea, opaca, prothoracis lateribus, abdomine, pedibusque in medio fuscescentibus; antennis tenuibus, elongatis, corpore parum brevioribus; prothorace subinæquali, crebre punctato, setulisque depressis concoloribus munito; elytris dense punctatis et setulosis, lineis elevatis longitudinalibus obsoletis. Long. corp. 9-11 mm.

Var. abdomine testaceo, concolore.

One example from each locality. Though differing in the colour of the hind-body beneath, I have no doubt they are one species; the specimen from Machachi with dark hind-body being the male.

COLEOPTERA—(Continued).

By the Rev. HENRY S. GORHAM, F.Z.S.

Fam. *ELATERIDÆ*.

1. *Athöus dispar*, n. sp.

Hab. Eastern side of Chimborazo (11,700 feet). Four specimens.

Nigro-piceus, nitidus, antennis, palpis, pedibusque rufis, capite crebre, prothorace minus crebre distincte punctatis; elytris punctato-striatis, punctis obsoletis. Long. 13-17 millim. ♂ ♀. Mas, minor, prothorace oblongo lateribus leviter sinuatis.

Femina, latior, prothorace latiori magis convexo lateribus fortius sinuatis, elytris convexioribus pone medium leviter ampliatis.

Head with the nasal portion well produced, coarsely punctured, raised at the edge and a little corrugated; frontal ridge distinct, flattened between the antennæ and more finely punctured. Antennæ (of the male) about two-thirds as long as the body, basal joint a little longer than the fourth, second joint equal to the third, and these together shorter than fourth. The remainder subequal to the fourth joint with the exception that in one of the two male specimens the ninth and tenth joints are shortened and distorted; in the female the three basal joints bear the same proportion, but the remainder are all rather shorter. The thorax is closely but distinctly and deeply punctured, in that of the male there is scarcely room between the

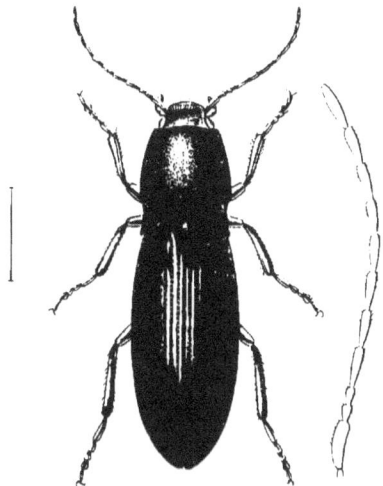

ATHOUS DISPAR, GORHAM.
CHIMBORAZO, 11,700 FEET.

separate punctures for another of the same size, but in the female there are here and there spaces which would admit one. The hind angles are produced, compressed and margined. Elytra faintly pubescent toward the apex. Underside rather evenly distinctly punctate. Mouth red, mandibles pitchy.

2. *Pyrophorus noctilucus*, Fab.

Hab. Nanegal (3-4000 feet). Two examples.

3. *Semiotus imperialis*, Guérin.

Hab. Nanegal (3-4000 feet). Six examples.

Fam. DASCILLIDÆ.

4. *Artematopus* sp.?

Hab. Forests above the Bridge of Chimbo (1-3000 feet). One example.

Agrees with a specimen in my own collection from Rio Janeiro. The species of this genus would therefore appear to have a wide range, the present one is a brown insect seven millimètres in length with punctate-striate elytra, and the antennæ longer than the whole body. The whole insect is sparingly pubescent including the antennæ. It is allied to *A. longicornis*, Perty.

5. *Ptilodactyla* sp. ?

Hab. First camp on Pichincha (14,000 feet). A single specimen.

One of the larger species, seven and a half millimètres in length, chestnut-brown, with obsoletely striate elytra, and ferruginous legs and antennæ.

Fam. *LYCIDÆ.*

The genus *Calopteron* includes numerous species from both North and South America, but is not found out of the New World. The first section of these have the elytra wonderfully dilated posteriorly, and often very beautifully reticulated, of a metallic blue colour, with white or yellow fasciæ or marks. The single specimen collected by Mr. Whymper belongs to this section, but is of moderate size and development compared with many species from lower regions.

6. *Calopteron Steinheili*, Bourgeois, Ann. Soc. Ent. Fr., 1878, p. 168. Cat. d. Lyc. rec. par M. Ed. Steinheil, Ann. Soc. Ent. Fr., 1879, Pl. 4, f. 7.

Hab. Valley of Collanes, Altar (12,500 feet). One specimen.

Three species of *Calopteron* are described by Bourgeois from Colombia which have the head, thorax, elytra, body, and legs, black or bluish black, with the exception of a marking near the apex of the elytra which is yellow. Of these the specimen captured by Mr. Whymper appears to me both from the description and the figure to agree most nearly with the species to which it is here referred, it is, however, very nearly related to *C. Poweri.*

7. *Plateros? alticola*, n. sp.

Hab. Hacienda of Guachala (9217 feet). A single specimen.

Oblongus subparallelus, niger, opacus, thorace parum nitido, antice profunde punctato, medio sulcato, lateribus elevatis, parallelis, angulis posticis acutis, prominulis ; elytris decem-striatis, interstitiis alternis elevatioribus, squamosis, ochraceis, basi circum scutello, sutura apiceque sat late nigris, antennis corporis dimidio parum longioribus ; articulo tertio, secundo sesqui longiore, subquadrato, articulis quarto ad undecimum subequalibus, haud serratis. Long. 8 millim. ♂ ?

This species, of which also only one specimen was secured, has the general

APPENDIX—COLEOPTERA. 47

structure and appearance of a species of the genus *Plateros*, the elytra in their close striation, punctuation, and scale-like but very close vestiture entirely agree with that genus, the antennæ also agree, the head is a little more prominent, but the prothorax is unlike that of *Plateros*, the sides being rather contracted below the middle, and the hind angles turned outward, the sides and middle of the disk are raised, thus leaving a deep and wide fovea on each side while the centre itself has a deep rather rudely-formed channel, not open at the base nor reaching within one-third of the front.

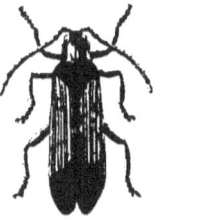

PLATEROS? ALTICOLA, GORHAM.
HACIENDA OF GUACHALA.

The characters which separate genera of *Lycidæ* are so difficult to seize and define, that I would not undertake to institute a genus upon a single specimen.

Fam. LAMPYRIDÆ.

Cladodes, Solier, is a genus of *Lampyridæ*, of which, generally, the species are of large size, with antennæ flabellate on one side, and with a very small

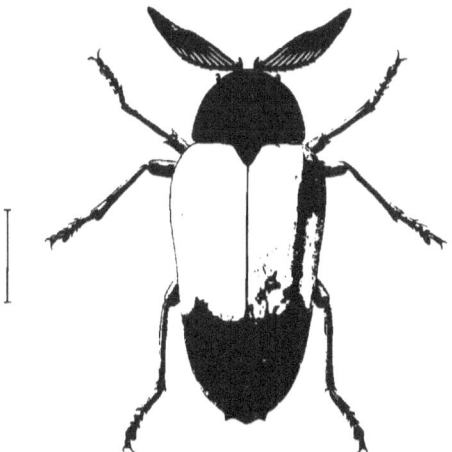

CLADODES NIGRICOLLIS, GORHAM.
HACIENDA OF GUACHALA.

portion of the abdomen capable of transmitting light. It is remarkable that hitherto we have been unable to distinguish the sexes, probably all the specimens sent are males, the female remaining unknown owing to some peculiarity in habit.

8. *Cladodes nigricollis*, Gorham, Trans. Ent. Soc., 1880, p. 8.

Hab. Hacienda of Guachala (9217 feet). Three examples.

Ater, opacus, thorace brevi, semicirculari, disco subdepresso; elytris croceis, postice attenuatis, triente apicali nigro, tarsis subtus griseo-pubescentibus, unguibus rufis. Long. 15 millim. ♂. See Figure on p. 47.

Of this species I have before only seen one specimen—the type which is now in the Brussels Museum and was obtained in Ecuador; and I can only identify it by memory and from my description. I have therefore made a rather longer diagnosis from the three specimens taken by Mr. Whymper. Not having before seen *Lampyridæ* from such a high elevation, it is a most interesting addition to our knowledge of this genus. This species is not very nearly allied to any of its congeners, but is nearest to a Chilian species.

9. *Photinus longipennis*, Mots. Etud. Ent., ii, p. 37; Gorh., Trans. Ent. Soc., 1880, p. 24.

Hab. Corredor Machai, Sara-urcu (12,700 feet) ♂, and La Dormida, Cayambe (11,800 feet) ♀. Two specimens.

A common Colombian insect. The females have shortened elytra, and are much rarer in collections than the males.

Photinus, Castelnau, is a New World genus and evidently one of large extent. The Central American species enumerated in Godman and Salvin's "Biologia" alone reach about fifty, but this number will convey a very imperfect idea of the Tropical South American species, when they shall be worked up. They are known by their ovate form, simple antennæ, and often highly luminous abdominal segments. The species recorded here is one of the largest.

10. *Cratomorphus discornfus*, Kirsch, Berl. Ent. Zeit., 1865, p. 72.

Hab. Nanegal (3-4000 feet). One example. Previously obtained in Colombia.

The character which distinguishes the specimen obtained by Mr. Whymper, and which is a male, is that the seventh ventral segment is excavated on its apical margin, disclosing a short small eighth segment, and the pygidium broadly truncate very faintly bisinuate. It is undoubtedly very near the species which I have identified with *C. fuscipennis*, Mots., and I have received that species from Peru. It is, however, rather more contracted towards the apex of the elytra than is usual in that species.

Cratomorphus, Motschulsky. Includes some of the largest known *Lampyridæ*. The males have very large eyes, which constitute by far the larger portion of the head. The abdominal ventral plates are different in the sexes and afford the best differential characters for species; these are, however, often very hard to discriminate, and I have some hesitation as to the species to which Mr. Whymper's specimen ought to be referred.

11. *Phengodes pulchella*, Guérin, Rev. Zool., 1843, p. 17; Lacord., Gen. des Col. Atlas, Tab. 44, f. 6.

Hab. Chillo (9000 feet). One example. Previously obtained in Colombia.

A single specimen taken by Mr. Whymper seems so close to this species that I cannot venture on separating it, but it differs from a specimen in my own collection in having the head and thorax nearly smooth. The sixth and seventh segments are brilliant ivory white and shining, yet it is not known that these insects are luminous. I think it probable that they are in some way parasitical upon true *Lampyridæ*.

Fam. TELEPHORIDÆ.

12. *Telephorus monticola*, n. sp.

Hab. Between Machachi and Pedregal (10,000 feet). A single example.

Niger, opacus, thorace nitido albo, disco nigro, ad angulos anticos albo; capite sublævi occipite subcarinato, ore et ad antennarum insertionem albido, oculis valde prominentibus, antennis nigrofuscis, articulo secundo tertio plus quam duplo breviore, articulorum apicibus angustissime albis; pedibus sat longis, nigrofuscis, coxis et trochanteribus dilutioribus; elytris longis ad apicem amplioribus, opacis crebre subtilissime rugosis. Long. 12 millim. ♂.

Mas, unguiculis anterioribus fissis, segmento septimo ventrali fisso.

This is a species at first sight having altogether the appearance of a *Podabrus*, the head is contracted behind the prominent eyes in the same manner, the thorax is small and short with its margins gently reflexed, but the structure of the claws is that of the species of the Central American genus *Discodon*, Gorham, and so is the form of the seventh ventral segment, which is divided by a fine suture for its entire length. The anterior claws of the front tarsi have a lamina, are more bent than the posterior ones of the same foot. The anterior ones of the intermediate and hind feet appear cleft. From typical *Discodons* this insect differs in not having the edge of the prothorax nicked. In most *Podabri* the third joint of the antennæ is scarcely longer than the second, here it is more than twice as long.

The position, therefore, of this species must remain conjectural till more specimens have been seen, and of both sexes.

13. Xenismus Whymperi, n. sp.

Hab. Ecuador (locality unknown).[1] Two specimens.

Testaceus, capite nigro glabro, antennis palpisque ferrugineis, his apice, illis articulis singulis apicibus summis nigris; prothorace subquadrato leviter transverso, glabro, nitidissimo, in medio signaturâ irregulari nigro; elytris pallide flavis, postici attenuati, basi (margine reflexo praetermisso) nigris; pectore fusco, pedibus ferrugineis, abdomine testaceo lateribus infuscatis. Long. 12-13 millim. ♂ ♀.

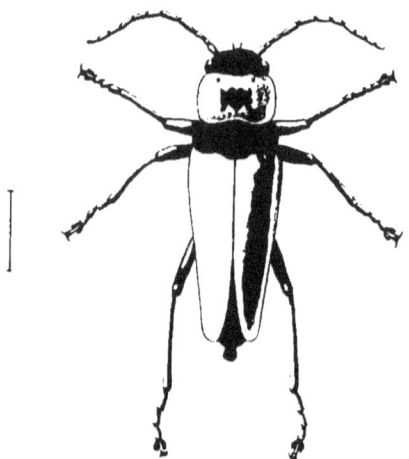

XENISMUS WHYMPERI, GORHAM.

Mas, segmento septimo ventrali (ut in *Chauliognatho*) convexo, elongato, genitali obtegente.

Femina, segmento sexto ventrali exciso, complicato, lobo laterali utrinque triangulari.

The genus *Xenismus* was proposed by Mr. C. O. Waterhouse in the Transactions of the Entomological Society for 1878, p. 331, for a species somewhat similar in colour to *X. Whymperi*, but with much more of the elytra black at the base and roughened (*X. nigroplagiatus*). Another species is known to me (*X. lusalis*, Dej.), figured by Lacordaire in the Atlas to his work on the Genera of Coleoptera (Tab. 45, f. 2); both these are from

[1] The locality was lost by the setter.—*E. W.*

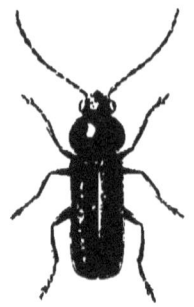

QUITO AND GUACHALA. 9 - 10,000 FEET

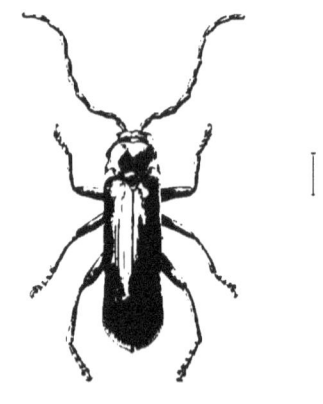

SOUTH SIDE OF CHIMBORAZO 15,000 FEET

Colombia or Ecuador, but are not generally known insects, and it is therefore highly interesting to record a third species. It is easily known from either of its congeners by the mark on the thorax, which is apparently variable but shaped something like the letter M with the central V filled up with black.

The genus is evidently allied to *Chauliognathus*, the head is wider in front and with a distinct clypeus. The claw joint of the tarsi is thickened in a club-like node at the insertion of the claws which are simple in both sexes. The black at the base of the elytra extends further in the female than in the male.

14. **Silis Chimborazona**, n. sp.

Hab. Southern side of Chimborazo, below the second camp (15,000 feet). One specimen.

Nigra, nitida, elongata, antennis fere simplibus, ore, mandibulis (apice excepto) abdominis apice, prothoraceque flavo-ferrugineo, hoc margine antico, discoque postice nigris. Long. 6½ millim. ♂.

Mas, prothoracis lateribus biexcisis, inter excisiones acute retrorsum productis, et spinâ nigrâ munitis; segmento septimo ventrali fisso, flavo.

This is a most interesting and remarkable insect allied to such species as *Silis dilacerata*, Gorh., Biol. Centr. Am., Vol. iii., pt. 2, p. 96, but differing from anything previously known in the mode of distortion of the lateral margin of the thorax into points and spines. The anterior black margin is thickened near the front angles and abruptly terminated there, below it springs the production of the middle of the sides—this part is red and has a sharp point of the same colour, but gives rise to another longer black spine which is hairy. Below it the margin appears excised, the base is margined. The elytra are black faintly substriate, opaque. For Figure see the accompanying Plate.

Silis is a genus comprising some of the most curious forms among the *Telephoridæ*, the prothorax having its margin distorted and cut away at or near the base, and often furnished with a spine-like production. The seventh ventral plate is split and covers the eighth, the claws are simple in the males.

PLECTONOTUM, nov. gen.

Corpus parvum; antennæ simplices, undecim-articulatæ corporis longitudine. Palpi articulo ultimo ovoido apice acuminati. Pronotum transversum, disco convexo, medio leviter canaliculato, margine parum elevato, margine laterali subincrassato, ad angulos posticos leviter exciso. Pedes sat longi. Elytra ampla, abdominis apicem vix tegentia.

This genus has at first sight very much the appearance of *Attalus* or *Anthocomus*, but it evidently belongs to the Telephoridæ, and is, I think, near *Silis*. I have not seen any species of Motschulsky's genus *Malthesis*, but it is very possibly allied to this genus. The form of the last abdominal segments precludes the idea that they can be congeneric.

I have had to use this generic name for a species at first sight very similar to the present one, collected in numbers by Mr. Champion at 3-4000 feet altitude on the Volcan de Chiriqui in Panama, and which will be described shortly in the Biologia Centrali-Americana as *Plectonotum labiale*. A closely allied form at present not described occurs in New Zealand. It appears to be a very primitive Siloid form.

15. *Plectonotum nigrum*, n. sp.

Hab. The Panecillo, Quito (10,000 feet), Hacienda of Guachala (9217 feet). Four specimens.

Nigrum, subnitidum, capite prothoraceque obsolete punctatis, hoc transverso, margine subelevato, lateribus in medio subincrassatis, angulis anterioribus obsoletis, posterioribus acutis, parum prominulis. Corpore subtus tibiisque infuscatis, antennis articulis duobus basalibus ferrugineis. Long. 3-3½ millim. ♂ ♀.

Mas, oculis magis prominulis abdominis apice albo, trimucronato.

The chief peculiarity of this little *Telephorid* consists in the curiously reflexed lateral margins of the thorax. The abdominal structure is not well observed, there being but a single male, the dorsal part of the genitalia appears to be bimucronate, while the acuminate ventral portion curves upwards between these two mucros. For Figure see the Plate facing p. 51.

Fam. MELYRIDÆ.

16. *Astylus bis-sexguttatus*, n. sp.

Hab. Type form, Quito to Guallabamba (9000 feet), Hacienda of Guachala (9217 feet), Pichincha (11-12,000 feet), Cotocachi (11-13,500 feet), Antisanilla to Piñantura (11,000 feet), Machachi (9-10,000 feet), Quito (9500 feet).

Var. a. Eastern side of Corazon (12,000 feet).

Var. b. Hacienda of Guachala (9217 feet). Very numerous specimens.

Niger, subnitidus, dense atripilosus, prothorace crebre haud distincte, elytris parcius profunde punctatis, his singulatim guttis sex rubris notatis, tribus subapicalibus sæpe confluentibus annulum interdum formantibus; antennis articulis tribus vel quatuor basalibus saturate rufis. Long. 7½-8 millim.

Var. a. Guttâ subapicali deficiente.

Var. β. Nigrâ elytris maculâ parva basali parum distinctâ.

This *Astylus* has been in my collection for some time, having been previously sent from Ecuador, but I have not been able to identify it with any described species. It is allied to *A. rubripennis* and *A. Bonplandi*, but its elytra are very much more rugosely punctate than in either of those species. It is also smaller. It appears to be very variable, the specimens previously sent from Ecuador have a complete but irregular annulus near the apex of the elytra, enclosing a somewhat triangular black spot. In those collected by Mr. Whymper the spots are not of a blood-red colour but of an orange yellow shade, and the annulus is seen to consist of three spots often disconnected, while in several all the spots vanish with the exception of the basal one.

Mr. Whymper, however, informs me that when the insect is alive the spots are more distinctly red than they are now, the colouring having undergone alteration since the specimens have been in spirit ; and he says that he

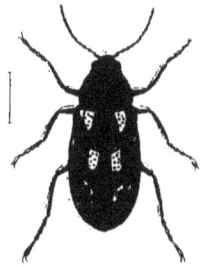

ASTYLUS BIS-SEXGUTTATUS, GORHAM.
9000-13,500 FEET.

"did not notice any other beetle in the interior of Ecuador which was so widely distributed, and in such large numbers. It was found almost everywhere, between 8-11,000 feet, upon trees, bushes, and plants, and was conspicuously numerous."

Astylus is a genus highly characteristic of the South American tropical fauna. They are hairy insects, representative of the Old World *Dasytes*, and having similar habits, sitting on plants and in flowers in the hot sunshine. About twenty species are now described.

17. *Listrus cœnescens*, n. sp.

Hab. The Panecillo, Quito (10,000 feet). A single example.

Niger, punctatus, elytris subæneis, antennis rufis, articulis quatuor apicalibus nigris, pedibus obscure rufis, femoribus piceis, capite crebre et distincte, prothorace parcius punctatis, hoc alutaceo, illo inter antennas bifoveolato ; elytris crebre punctatis. Long. 3½ millim.

Head with the eyes moderately prominent ; antennæ longer than the head and thorax, basal joint red, spotted with black above, the three or four apical joints scarcely form a club, but are wider and black ; thorax not wider than longer, but a little narrowed in front, margined by a fine line on the sides and base, and very indistinctly so in front. There is a faint shallow depression on the disk near the base, otherwise it is even, with distinct punctures. Elytra rather more shining than the thorax, closely and subconfluently punctate.

Listrus is a genus of small beetles allied to the European *Donacœa*. In the Biologia Centrali Americana eight species are described from that region, and about ten others are known, all North American. Kirsch's genus *Haplomaurus*, of which he has described two species from Bogota, seems to differ from this in having erect hairs and to be nearer to *Pristoscelis*.

18. *Listrus flavipennis*, n. sp.

Hab. Machachi to Pedregal (10,000 feet). Five examples.

Nigro æneus, antennis pedibusque ferrugineis elytris, flavo-testaceis, suturâ et circum scutellum ænescentibus, capite prothoraceque distincte parcius punctatis subtilissime alutaceis, elytris obsoletius punctatis. Long. 4-4¼ millim.

Head and thorax similar in form and in punctuation to *L. ænescens*; antennae with the two basal joints spotted with pitchy, but with that exception entirely yellow; palpi dark, elytra yellow, closely and rather obsoletely punctate, especially so near the base and suture, the latter pitchy, or black and with a black patch round the scutellum; the shoulder is rather tubercular. Legs chestnut yellow, tarsi of the same colour with the rest of the legs.

Fam. PTINIDÆ.

19. *Trigonogenius squalidus*, Boield, Mon. Ptin., p. 667; Chevr. Ann. Fr., 1861, p. 389.

Hab. Machachi (9-10,000 feet). A single example.

This is a small almost globular-bodied *Ptinus*, with a broad thorax, and covered densely with grey brown scales, and erect black hairs on the thorax and elytra.

Very few *Ptinidæ* have at present been recorded from the South American continent, this species with one other (*T. globulum*) being all that have yet been observed, with the exception that Chili has three or four.

Trigonogenius is a Chilian type, with which have been associated some species from the Atlantic Islands, and by myself a species from Central America. The species to which we refer Mr. Whymper's Ecuador insect has been recorded from Colombia.

Fam. HISPIDÆ.

20. *Aroscus parumpunctatus*, n. sp.

Hab. Nanegal (3-4000 feet). A single specimen.

Niger, nitidus capite cornu luto subquadrato basi parum ampliato, prothorace transverso, angulis anticis depressis, subrectis, posticis acutiusculis, disco convexo, basi sinuato in medio depresso; scutello nigro. Elytris luteis obsolete punctato-striatis, striis extrorsum et ad apicem obliteratis; punctis tribus nigris, una humerali una discoidali ante medium, una majore pone medium lunata. Long. 13 millim.

Apparently allied to *A. perplexus*, Baly, but apart from the colour (upon which in this variable genus much stress cannot be laid) the elytra are not angularly produced, their apex only being slightly emarginate; nor are they distinctly striate, but have rows of close, small brown punctures, interspersed with irregular ones at the sides.

Arescus is one of the most beautiful and interesting genera of the *Hispidæ*, combining strangeness of form with beauty of colour. Six species are described, all from South America.

Fam. CASSIDIDÆ.

21. *Physonota dilatata*, Kirsch, Deuts. Ent. Zeits., 1876, p. 93?

Hab. Nanegal (3-4000 feet). Two specimens.

These appear to me to agree fairly with the description given by Kirsch of his species from Chanchamayo. The very large size of one of these (19 millimètres in length), the colour of the antennæ, of the underside, and elytra, seem to point to the conclusion that our species is identical. A second obviously conspecific measures 15 millimètres. In neither is the suture black at the apex beneath, and some minor discrepancies such as may be due to variation are observable.

22. *Chelymorpha* sp.?

Hab. Banks of the Guayas, Guayaquil. Two specimens.

The species of this genus are difficult to identify. The specimens, both from Guayaquil, probably belong to different species, but are closely allied in form and sculpture, and pertain to the section of the genus with the elytra not gibbous. In one, the elytra have numerous small black spots, and the body is black beneath; while in the other, the elytra are ferruginous, with only one or two spots near their base, and the body is of the same ferruginous colour, as well as the legs.

23. *Coptocycla* sp.?

Hab. Banks of the Guayas, Guayaquil.

Three specimens of a *Coptocycla* of the size of, and in its gibbous-convex form allied to, *C. judaica*, but differing in colour, having the disk of the elytra and of the prothorax deep pitchy brown. These specimens I cannot identify with any described species at present, but it is hardly advisable to give them a name.

Mr. Whymper informs me that his specimens were amongst stones and short grass on the banks of the Guayas just outside Guayaquil. When living they were of a golden colour.

Fam. EROTYLIDÆ.

24. *Morphoides hæmatocephalus*, Lac. Mon., p. 361. *Var.*

Hab. Banks of the Guayas, Guayaquil; Nanegal (3-4000 feet). Four specimens.

The specimens which I refer to a variety of this species are a trifle more oval than usual, and have the base of the head black. It is noticeable that though *M. hæmatocephalus* is variable in the length of the elytral black patch, in the depth of the striation, and slightly in form, I have not before found specimens with any tendency to vary in the head.

Morphoides is in Crotch's estimation simply a section of the great genus *Brachysphœnus*. The section, if restricted, consists of species having a strong general resemblance. The elytra are red, generally with a black central patch, and the head is often red, the thorax black or reddish, but more often the former. *M. bimaculatus*, Germ., is the type of this section.

25. *Homoiotelus acuminatus*, n. sp.

Hab. Nanegal (3-4000 feet). Two specimens.

Ovatus, antice et postice acuminatus, testaceus antennis (articulis duobus basalibus praetermissis), scutello, pectore, abdomine, femoribus apice, tibiis tarsisque nigris, elytris obsoletius gemellato, punctato-striatis. Long. 9-10 millim.

The colour of this species is castaneous, the head and thorax smooth, the latter trapezoidal, with the sides strongly narrowed in front, and the base a little rounded. The scutellum is black, but the mesosternum is yellow, the epimera and metasternum with the abdomen quite black. The series of punctures are distinct at the base, but become very obsolete about the middle, and vanish before the apex of the elytra. The black parts of the body beneath distinguish this from *O. gemellatus*; the spotless thorax, colour, and small size from *O. crocicollis*. It seems abundantly distinct from any described species.

26. *Ægithus truncatus*, Crotch, Revision of Erotylidæ, p. 116; Cist. Ent., xiii, 1876, p. 492.

Hab. Tanti (1890 feet). Previously obtained in Ecuador by Mr. Buckley.

There is a single specimen of an *Ægithus* in Mr. Whymper's collection which I refer to this with very little doubt. It is easily recognisable by its being very convex, the head and thorax black, the former obscurely red beneath, the latter pitchy at the sides, the elytra smooth and polished without striæ or punctuation.

27. *Ægithus uva*, Lac., Mon. Erotyl., p. 290.

Hab. Nanegal (3-4000 feet).

The only difference between one specimen of this and my examples of *Æ. uva* from Colombia is that the abdomen in this one is yellowish. I do not think, however, it is distinct.

Fam. COCCINELLIDÆ.

28. *Eriopis connexa* (*Coccinella connexa*), Germar, Ins. sp. nov., p. 62.

Hab. Hacienda of Guachala (9217 feet), Cayambe village (9220 feet), Machachi (9-10,000 feet). Nine specimens.

Some species of *Coccinellidæ* have a remarkably wide range of distribution, and sometimes with less variation than would naturally be expected under such circumstances. The present species occurs from the Straits of Magellan to Vancouver's Island, and apparently is not restricted to the Pacific side of the Southern Continent, being recorded from Brazil, Montevideo, etc. It is variable, more so than usual in this family.

29. *Megilla maculata* (*Coccinella maculata*), De Geer, Mim., v, p. 392 ; T. 16, f. 22.

Hab. Bodegas (level of sea). Three specimens.

Very variable in size and somewhat so in the ground colour, and size of spots. This has almost as extended a range as *E. connexa*. It is recorded from Chili, and as far north as Canada. It is common in North America. In the Amazon district a large and developed variety occurs.

30. *Neda Norrisii* (*Coccinella Norrisii*), Guérin, Icon. Regn. An., p. 320.

Hab. Nanegal (3-4000 feet). Numerous specimens.

A common species in Ecuador, occurring also in Colombia. The elytra vary from pale straw-colour to rich red, and have usually four black squarish spots—one basal, one transverse marginal with one opposite it near the suture, one submarginal, one third from the apex, but there is a variety in which the basal spot is wanting.

31. *Cycloneda* sp.?

Hab. Eastern side of Corazon (12,000 feet). One example. Quito (Coll. Gorham, A. Murray).

This is a small species barely exceeding 4 millimetres in length, black, the head with two white spots, the thorax margined neatly with white, and with two linear white spots; the elytra orange with the basal margin pale. I am not able to ascertain if it has been described.

Cyclonoda was proposed by Crotch for the species placed under *Daulis* by Mulsant, with some few others. They are all American, the type *C.* (*Daulis*) *sanguinea*, L., being a widely spread and very abundant species with un-spotted red elytra.

32. *Scymnus* sp.?

Hab. The Panecillo, Quito (10,000 feet). One specimen.

A single example of a small yellow *Scymnus*, which is probably undescribed, but which is not in a condition to allow of its being characterised, was met with. The genus is pretty nearly if not quite cosmopolitan.

COLEOPTERA—(Continued).

By A. SIDNEY OLLIFF.

CLAVICORNIA.

The two *Nitidulidæ* obtained by Mr. Whymper appear to be undescribed. For one of them I am constrained to establish a new genus, as it differs considerably from anything previously known. The family *Trogositidæ* is only represented by the widely distributed *Trogosita cænea* which was described from Brazil, and is also found at Porto Rico and Jamaica.

Fam. *NITIDULIDÆ.*

1. *Cercometes Andicola*, sp. n.

Hab. Eastern side of Corazon (12,000 feet).

Elongate-ovate, moderately convex, bright metallic green, shining, head and prothorax very slightly darker. Head transverse, closely and very finely punctured on the crown, less closely punctured at the base. Antennæ fulvous, the club slightly darker. Prothorax rather convex, considerably wider than long, narrower in front than behind; the disc extremely finely coriaceous and very finely and not very closely punctured, the sides and shoulders under a moderately strong magnifying power appear rather more closely punctured and coriaceous; sides arcuate, very slightly sinuate just before the basal angles which are obtuse. Scutellum broader than long, pointed behind, regularly and very delicately punctured, the sides arcuate. Elytra rather longer than wide, obliquely truncated behind, moderately convex, extremely finely asperate-punctate; the humeral angles not very prominent; the sides regularly arcuate; the outer apical angles rounded, the sutural angles slightly obtuse. Pygidium very finely and closely punctured, sparingly covered with very fine yellowish grey pubescence. Underside not quite such a bright metallic colour as above; prosternum finely coriaceous; meso- and metasternum extremely finely punctured. Legs reddish testaceous, claws fuscous. Length, 2½ mm.

This species is allied to *Cercometes politus*, Reitter (Verh. des naturf. Ver. Brünn, xii, p. 167, 1873), from Colombia and *C. Deyrollei*, Reitter (l.c. xiii, p. 100, 1875), from Brazil. Judging from the description, it may be distinguished from the former by its being of a metallic green colour instead of dark black, and in having the posterior angles of the prothorax obtusely rounded and not right angles; and from the latter, not only in colour but also in having the prothorax finely punctured and coriaceous on the disc. At first sight it somewhat resembles *Cercus abdominalis*, Erich., but structurally is quite different; the claws are strongly toothed at the base, and the labial palpi appear to be four-jointed, thus agreeing with the genus *Cercometes* of Reitter.

PLEURONECES, gen. nov.

Body depressed and rather broad. Head moderately large, transverse. Eyes rather large, prominent and not very strongly facetted. Antennæ eleven-jointed, the basal joint rather large, the second to fifth narrower, rather shorter, and of about equal length, sixth to eighth much shorter, terminated by a tolerably compact club of three joints. Labrum strongly bilobed. Mandibles strongly recurved, the apex acute, with four or five teeth on the inner margin, the last of these teeth rather larger than the others. Maxillæ with the lobe enlarged in front, the fringe long and somewhat bristly, only extending on the inner margin for about one-third of its length; maxillary palpi four-jointed, the basal joint moderately large, the second slightly longer and much narrower at the base than at the apex, the third decidedly shorter, the apical longer than the second and third together, rounded at the extremity. Labrum small; the labial palpi appear to be three-jointed, the first very small, the second much longer and broader, the terminal joint as long and rather broader than the second. Prothorax transverse, narrowly margined, free at the base and not overlapping the elytra. Elytra truncate behind, leaving the pygidium wholly exposed. Prosternum rounded behind the coxæ. Metasternal episternum distinctly pointed behind.

First abdominal segment slightly longer than the four following which are almost equal in length. Legs moderately robust, tarsi dilated, claws feebly dilated near the base.

The males with an accessory segment visible from above.

After considerable hesitation I have come to the conclusion that this genus is best placed near *Nitidula* and *Epuræa*. The dentate mandibles, sharply pointed metasternal episternum, and the sexual character will suffice to distinguish it from these and the allied genera.

2. *Pleuronecces montanus*, sp. n.

Hab. Antisanilla to Piñantura (11,000 feet).

Elongate, depressed, slightly broader behind than in front, black, shining, the elytra rather dark brownish testaceous. Head moderately strongly and not very closely punctured; the epistoma only punctured at the base. Antennæ reddish testaceous, the club pitchy black. Prothorax much broader than long, moderately convex, finely margined, as strongly and about as closely punctured on the disc as the head, the punctures slightly closer near the sides; all the angles obtuse; the sides arcuate. Scutellum large, finely punctured at the base, the sides oblique and slightly arcuate, the apex rounded. Elytra about one-third longer than broad, truncate behind, not very convex, at the base as broad as the prothorax, slightly broader posteriorly, finely margined at the sides, rather strongly and closely punctured; humeral angles rather prominent; outer apical angles strongly and sutural angles feebly rounded. Pygidium densely but not very strongly punctured, sparingly clothed with short bristly pubescence. Underside pitchy black; prosternum finely punctured and coriaceous; mesosternum, metasternum and abdominal segments rather more strongly and regularly punctured. Legs brownish testaceous, femora somewhat darker. Length, $4\frac{1}{2}$ mm.; greatest width, $2\frac{1}{4}$ mm. For Figure see the accompanying Plate.

Fam. TROGOSITIDÆ.

3. *Trogosita* (*Temnochila*) *ænea*, Oliv., Ent. ii, 19, p. 7, pl. 1, fig. 3 (1790); Reitter, Verh. des naturf. Ver. Brünn, xiii, p. 14 (1875).

Hab. Pacific slopes (below 1400 feet). A single example.

RHYNCHOPHORA.

Amongst the Rhynchophora in Mr. Whymper's Ecuadorian collections, the two large families *Otiorrhynchidæ* and *Curculionidæ* are, as might be expected, better represented than any others, twenty-eight of the thirty-four species

LIBRARY OF
U. S. GRANT

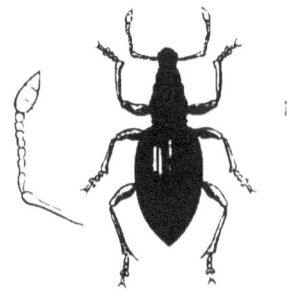

HELICORRHYNCHUS VULSUS, CLIFF
PICHINCHA & CHIMBORAZO, 15,500-16,000 FEET

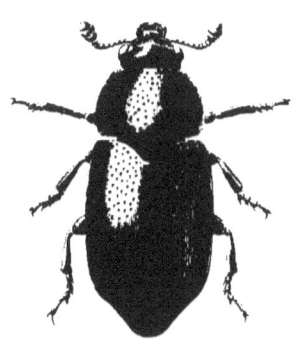

BETWEEN ANTISANILLA & PINANTURA, 11,000 FEET

obtained are referable to these families, whilst four pertain to the *Calandridæ* in its broadest sense. The *Brenthidæ* are represented by two species and were found at comparatively low altitudes. So many of the Rhynchophora are apterous that it would be unsafe to lay any great stress on the fact that a certain proportion of the species in the present collection are wingless; it may, however, be worth mentioning that *Macrops colorum* is apterous, as it was found at a very great elevation and belongs to a genus which has hitherto contained as far as I am aware only winged species.

The measurements in the following descriptions are *exclusive* of the rostrum except when the contrary is stated.

I have to thank Mr. F. P. Pascoe for kindly permitting me to see the types in his collection that I have found it advisable to examine.

The following are the new species described in this paper:—

Helicorrhynchus vulsus.
Pandeletius argentatus.
Compsus Whymperi.
Naupactus segnipes.
N. pauper.
N. nigrans.
N. parvicollis.
Amphideritus brevis.
A. pygmæus.
Listroderes inconspicuus.
L. punctatissimus.
Amathynetes alticola.

A. simulans.
Macrops colorum.
Anchonus monticola.
A. Altarensis.
Hilipus longicollis.
Erirrhinoides distinctus.
Erirrhinus glaber.
Otidocephalus? spinicollis.
Apion Andinum.
Sphenophorus notandus.
Cossonus coloratus.

Fam. OTIORRHYNCHIDÆ.

HELICORRHYNCHUS, gen. nov.

General characters of *Otiorrhynchus*; rostrum stout, longer than the head, the tip feebly emarginate in the middle, with a slight sinuation on each side near the anterior angles and a moderately strong circular impression on the disc between the antennæ. The antennæ long; scape slightly sinuous in the middle; funiculus seven-jointed, the first two joints only slightly longer than the following ones; club elongate. Scrobes deep, lateral, strongly arcuate and almost reaching the lower margin of the eye. Eyes moderately large, round, not very prominent. Prothorax with a very strong circular impression on each side behind the middle. Scutellum moderately large and distinct. Elytra elongate, oval. Metasternum rather short, the sidepieces very narrow. The third and fourth abdominal segments nearly equal in length, the two together about as long as the second segment. Legs long, femora clavate, tibiae very feebly arcuate.

This genus is allied to *Otiorrhynchus*, Germ., and *Sciopithes*, Horn (Proc. Amer. Phil. Soc., xv, p. 62, 1876), but differs in the structure of the rostrum and antennal scrobes, etc., and also from the latter in the presence of a distinct scutellum.

4. *Helicorrhynchus vulsus*, sp. n.

Hab. Pichincha (15,500 feet), Chimborazo (15,800-16,000 feet).

Elongate-ovate, moderately convex, pitchy black, shining. Head rather broad, finely and closely punctured, with an elongate feeble impression between the eyes and a much shorter and deeper impression at the base which is sometimes hidden beneath the prothorax; rostrum broader in front than behind, slightly constricted in the middle, more strongly punctured than the head, finely and irregularly strigose near the sides, with a moderately strong longitudinal impression near the apex between the antennae. Antennae ferruginous, clothed with fine yellow pubescence. Prothorax slightly longer than broad, narrower in front than behind, extremely finely and rather closely punctured, with a very strong circular impression on each side just behind the middle; no distinct median line; sides slightly constricted both before and behind, arcuate in the middle. Scutellum impunctate. Elytra at the base considerably broader than the prothorax, still broader behind the middle especially in the female, about twice as long as the head and prothorax together, moderately strongly punctate-striate, the striae effaced posteriorly and indistinct towards the sides, the interstices rather broad, flat, very finely aciculate and extremely finely punctured; the sides arcuately rounded to the apex. Underside pale pitchy, finely and not very closely punctured, very sparingly clothed with yellow pubescence. Legs ferruginous. This species has much the appearance of a small *Otiorrhynchus*. Length, 6-7½ mm. For Figure see the Plate facing p. 60.

5. *Pandeletius argentatus*, sp. n.

Hab. Eastern side of Corazon (12,000 feet).

Elongate, narrower in front than behind, rather convex, pitchy, densely clothed with metallic golden and silvery green scales. Head broad, moderately convex, slightly depressed in front, finely and not very closely punctured; eyes not very prominent; rostrum rather long, narrowed in front, obliquely depressed in the middle throughout its whole length, the central channel extending from between the eyes to just behind the apex, sparingly and finely punctured, the apex triangularly emarginate and clothed with rather long grey hairs. Antennæ pitchy red, the club darker and clothed with grey pubescence. Prothorax slightly broader than long, somewhat constricted both in front and behind, very sparingly and not very closely punctured; the sides regularly arcuate. Scutellum small, sides obliquely rounded, covered with dirty white scales. Elytra about one and a half times as long as the head (including the rostrum) and prothorax together, moderately strongly striate-punctate, the interstices broad and impunctate; the shoulders not very prominent; the sides arcuate. Underside pitchy black, very obsoletely punctured; the last two abdominal segments red, scaleless, covered with rather long grey pubescence. Legs pitchy red, the femora covered with scales: the anterior legs much longer than the others; the femora crassate, very broad, of a rather darker pitchy colour, with a strong tooth on the inner margin near the apex; the tibiæ with a smaller tooth near the base and a row of short sharp teeth on the inner margin towards the apical half. In the female the elytra are much broader behind than in the male.

A very distinct species allied to *Pandeletius tibialis*, Bohem., from Mexico; differs in colour, in having the anterior femora toothed and the prothorax broader.

Length, ♂ 6½; ♀ 7 mm.

6. *Polydacris*, sp.

A single imperfect specimen.

Hab. Camp at Corredor Machai, Sara-urcu (12,700 feet).

7. *Compsus Whymperi*, sp. n.

Hab. Ambato (8000 feet); Penipe to Riobamba (9000 feet).

Ovate, narrower in front than behind, convex, dull black, finely setose and sparingly covered with dirty white pubescence. Head smooth; rostrum with a moderately strong groove on each side extending from the apex nearly to the eyes and approximating behind; median line not very strongly marked. Antennæ covered with very fine dirty white scales and pubescence. Prothorax longer than broad, somewhat narrowed in front, rather strongly rugose, with two strong approximate ridges on the disc, the space between these ridges nearly smooth and impunctate; sides feebly arcuate. Scutellum very small. Elytra ovate, convex, at the base much broader than the prothorax, broader behind the middle, rather strongly and irregularly rugose-punctate, the depressions rather strong on the disc, stronger near the sides where they are like small pits, with a moderately strong irregular ridge on each side of the suture, usually in the third interstice, commencing at the base and disappearing behind the middle; humeral angles somewhat produced. Underside dull black, extremely finely rugose, very sparingly covered with fine setæ; the mesosternum rather more strongly rugose and covered with dirty white scales. Legs finely setose; femora black; tibiæ and tarsi covered with very fine dirty white scales. Length, 10-16 mm.

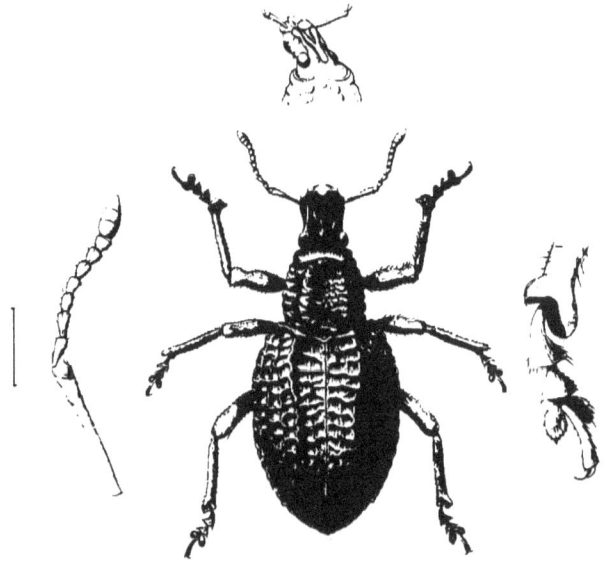

COMPSUS WHYMPERI, OLLIFF.
AMBATO, 8000 FEET.

Three specimens of this fine insect, differing somewhat among themselves but I believe representing one very variable species, were obtained by Mr. Whymper, to whom it is dedicated. The largest specimen, measuring 16 mm., differs from the other two in having the prothorax somewhat broader and the punctuation on the disc of the elytra more regular.

8. *Compsus* sp.

Hab. Pacific slopes (1-3000 feet).

A single malformed specimen of a species closely allied to *Compsus argyreus*, Linn., but having the prothorax longer and more densely punctured, and both the upper and under sides of the entire insect covered with silvery-grey and pale bluish-green scales. Length, 9 mm.

9. *Exorides carinatus*, Pascoe, Ann. Mag. Nat. Hist., (5) vii, p. 43 (1881).

Hab. Lower slopes of Antisana (13,000 feet). Six examples.

Appears to be a slight variety of this species, having the elytral carinae stronger than the type.

EXORIDES CARINATUS, PASCOE.
ANTISANA, 13,000 FEET.

10. *Praepodes annulonotatus*, Waterh., Cistula Ent., ii, p. 423, pl. ix, fig. 3 (1879).
Hab. Nanegal (3-4000 feet).

A small specimen which perfectly agrees with the type of this species in the British Museum from Medellin in Colombia. I have also seen this insect from the Balzar Mountains.

11. *Naupactus seynipes*, sp. n.
Hab. Machachi (9-10,000 feet).

Oblong-ovate, rather convex, much narrower in front than behind, black, densely clothed with brownish white scales. Head slightly narrowed in front, finely and rather closely punctured, the median line distinct, with a longitudinal depression extending from between the eyes to beyond the middle of the rostrum. Antennae piceous, the second joint of the funiculus only slightly longer than the first. Eyes rather large and prominent. Prothorax considerably broader than long, much narrower in front than behind, finely coriaceous, closely covered with small brown circular scales, with a feeble depression on each side near the base, median line nearly obsolete; sides strongly arcuate. Scutellum triangular, finely punctured and coriaceous. Elytra ample, about twice as long as the head and prothorax together, slightly narrower in front than behind, moderately strongly punctate-striate, the interstices rather broad, slightly raised and extremely finely coriaceous; humeral angles not very prominent; sides arcuately rounded to the apex. Underside black, sparingly covered with fine scales and pubescence, very finely punctured and coriaceous; mesosternum with a rather strong impression at

K

the base; metasternum with a much larger and more feeble impression in the middle. Legs black, tibiæ and tarsi piceous covered with brown scales; anterior tibiæ with a row of moderately strongly developed spines on the inner margin. Length, 10-11 mm.

Allied to *Naupactus ruricola*, Bohem., from Brazil; differs in being covered with brown instead of golden green scales, in having the prothorax proportionately broader, and the anterior tibiæ armed with short spines.

12. *N. pauper*, sp. n.

Hab. Quito (9500 feet); Pichincha (12-13,000 feet).

Elongate, rather narrow, moderately convex, piceous, closely covered with brown scales intermingled with a few paler silvery-green ones; the elytra pitchy and even more closely covered with brown scales. Head rather broad, extremely finely rugulose, with an indistinct median line; eyes large and prominent; rostrum somewhat constricted just beyond the middle, with a moderately strongly impressed longitudinal line in the middle extending throughout its whole length. Antennæ densely pubescent; the scape piceous, the funiculus and club ferruginous. Prothorax transverse, about as broad in front as behind, very finely punctured and rugulose, with an inconspicuous longitudinal line of green scales on each side of the middle; the anterior margin slightly raised and ferruginous; sides regularly arcuate; the median line very indistinct. Scutellum small, covered with white scales. Elytra more than twice as long as the head and prothorax together, considerably broader at the base than the prothorax, sparingly covered with short brown and rather bristly hairs especially on the disc, moderately strongly striate-punctate, the interstices rather broad and impunctate, the striæ more feeble near the apex; humeral angles slightly prominent; sides nearly parallel for two-thirds of their length, then arcuately narrowed to the apex. Underside piceous, thickly clothed with brown and silvery-green scales, extremely finely rugulose. Legs pitchy red, sparingly clothed with scales. Length, 7 mm. This species has something of the appearance of a *Sitones*.

13. *N. nigrans*, sp. n.

Hab. Between Quito and Guallabamba (9000 feet).

Elongate-ovate, rather convex, black, shining, sparingly covered with extremely fine yellow pubescence. Head finely and moderately closely punctured; eyes rather large, lateral and rather prominent; rostrum not very long, slightly narrower in front than behind, more strongly punctured than the head, the sides finely strigose; with a moderately strong longitudinal impression between the eyes extending from near the base of the head to just behind the apex of the rostrum. Antennæ pitchy, the club clothed with yellowish grey pubescence. Prothorax somewhat broader than long, narrower in front than behind, very sparingly and finely punctured; the anterior margin slightly thickened; the sides feebly arcuate. Scutellum small, triangular and impunctate. Elytra about as long as the head and prothorax together, slightly broader behind the middle than in front, the apical half sparingly clothed with rather long conspicuous yellow pubescence, moderately strongly striate-punctate, the punctures gradually disappearing posteriorly, the interstices broad, smooth and impunctate; humeral angles slightly prominent; the sides feebly arcuate and very slightly sinuous just behind the apex. Underside rather closely covered with fine yellowish grey pubescence, extremely finely and transversely strigose. Legs pitchy, rather densely clothed with long grey pubescence; the anterior tibiæ with a row of small teeth on the inner margin. Length, 11 mm.

Differs from all the species of the genus known to me in its short compact form, black colour, and glabrous appearance.

14. *N. parvicollis*, sp. n.

Hab. Cayambe (15,000 feet); Chimborazo (15,800 feet).

Elongate, narrower in front than behind, rather convex, piceous, shining, very sparingly covered with fine yellow pubescence. Head very broad, obsoletely and rather closely punctured, the median impression indistinct, the sides slightly swollen near the base; eyes lateral, not prominent; rostrum narrower in front than behind, very feebly constricted in the middle, with a strong longitudinal impression on the disc extending from between the eyes to just behind the apex. Antennæ reddish testaceous, the club dusky and clothed with short and rather dense grey pubescence. Prothorax very small, transverse, at the base as broad as the head, rather strongly constricted both in front and behind, finely and rather closely punctured in front, the punctures gradually disappearing posteriorly and leaving the basal half almost impunctate, the pubescence near the sides rather long and somewhat closer than on the disc; the lateral margins between the constrictions strongly arcuate. Scutellum small, nearly triangular, impunctate. Elytra more than twice as long as the head and prothorax together, ovate, narrower in front than behind, feebly striate-punctate, the striæ very obsolete posteriorly, the interstices broad, not raised and smooth; humeral angles rather prominent; sides arcuately rounded to the apex. Underside rather paler in colour than above; the sternum very finely punctured; abdominal segments finely aciculate transversely, very sparingly and obsoletely punctured. Legs pitchy red, the knees and upper sides of the femora near the base slightly darker. Length, 7½-9 mm.

Differs from all the described species of *Naupactus* in the large size of its head, the remarkably narrow prothorax and its comparatively smooth and highly polished upper surface. *N. gibbicollis*, Bohem., from Brazil, is black and highly polished, but has little in common with the insect characterised above.

15. *Amphideritus brevis*, sp. n.

Hab. Ecuador (no exact locality noted).[1]

Elongate-ovate, rather short, moderately convex, black, shining, sparingly covered with greyish white pubescence. Head finely punctured and pubescent; rostrum obliquely rugulose, with a moderately strong longitudinal impression extending from between the eyes to just behind the apex. Antennæ dark ferruginous, the club paler and closely covered with grey pubescence. Prothorax transverse, at its greatest width just behind the middle, rather narrower in front than behind, convex, strongly and rather closely punctured on the disc, near the sides somewhat more closely and less strongly punctured; without median line; the sides strongly arcuate. Scutellum very small, shining and impunctate. Elytra about twice as long as the head and prothorax together, strongly striate-punctate, clothed with fine brownish-white pubescence especially near the apex, the interstices moderately broad and smooth; shoulders not prominent; sides arcuately narrowed to the apex. Underside black, shining; the sterna finely and not very closely punctured; the first two abdominal segments more finely punctured, the others almost impunctate, shining and very sparingly pubescent. Legs finely pubescent; femora black; tibiæ and tarsi dark ferruginous. Length, 7½ mm.

Allied to *Amphideritus vilis*, Bohem., from Colombia, but is easily separated by its shorter and broader form, more strongly punctured prothorax, and the more strongly punctured striæ and less closely pubescent surface of the elytra.

16. *A. pygmæus*, sp. n.

Hab. Chimborazo (12-13,000 feet).

Elongate, rather convex, black, shining, sparingly and very finely pubescent. Head very finely punctured; rostrum finely rugulose, with a rather deeply impressed line extending from between the eyes to just before the apex. Prothorax very finely and sparingly punctured on the disc, the sides rather more strongly and much more closely punctured. Elytra punctured in rows, the punctures distinct near the base, gradually obliterated towards the apex, the interstices broad, flat and impunctate. Underside black, finely

---

[1] The locality was lost by the setter.—*E. W.*

pubescent; the sterna finely punctured; abdominal segments, except the last which is shining and impunctate, extremely finely punctured. Legs black and pubescent; tibiæ and tarsi dark ferruginous. Length, $5\frac{1}{4}$ mm.

Closely allied to the preceding species, but differs in being smaller and comparatively narrower, in having the prothorax much more finely and sparingly punctured and the elytra very feebly striate-punctate.

Fam. CURCULIONIDÆ.

17. *Listroderes inconspicuus*, sp. n.

Hab. Cayambe (15,000 feet).

Broadly ovate, narrower in front than behind, moderately convex, black, closely covered with short pale greenish grey decumbent pubescence. Head strongly convex, finely and closely punctured; rostrum with a distinct median ridge, the sides irregularly punctured and somewhat strigose. Antennæ pitchy red, finely pubescent. Prothorax strongly transverse, very slightly constricted both in front and behind, finely and very closely punctured; the sides arcuate. Scutellum very small, finely punctured. Elytra slightly more than twice as long as the head and prothorax together, moderately convex, densely pubescent, very feebly striate-punctate, the interstices broad; humeral angles not prominent; sides arcuately rounded to the apex. Underside coloured as above but much more sparingly pubescent; the sterna very finely and not very closely punctured; abdominal segments finely and closely punctured: the third and fourth segments shorter than the others but slightly longer than in the type of this genus. Legs black, the pubescence especially on the tibiæ inclining to yellowish. Length, 8 mm.

This species is best placed near *Listroderes subcostatus*, Waterh., described from Chili.

18. *L. punctatissimus*, sp. n.

Hab. Chimborazo (11,700 feet).

Broadly ovate, rather convex, pitchy black, somewhat shining, moderately closely covered with short erect pubescence; the elytra very densely covered with small sandy brown scales. Head broad, rather strongly and very closely punctured especially in front, with a small circular impression between the eyes; rostrum broader in front than behind, constricted just before the middle, even more closely punctured than the head, with an indistinct elevated line in the middle extending throughout its entire length. Antennæ pitchy red, clothed with very fine grey pubescence; the second joint of the funiculus rather shorter than the first. Prothorax transverse, slightly narrower in front than behind, rather strongly and extremely closely punctured, the pubescence more dense near the sides, with a small stripe of dirty white scales on each side above the posterior angle; median line not very distinct; the sides regularly arcuate. Scutellum small, rather finely punctured. Elytra about two and a half times as long as the head and prothorax together, moderately strongly punctate-striate; the striae rather obsolete, the punctures close, the interstices moderately broad, finely punctured and very slightly raised; densely covered with small sandy brown scales, with

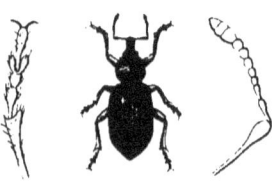

LISTRODERES PUNCTATISSIMUS, OLLIFF.
CHIMBORAZO, 11,700 FEET.

small white spots of scales placed at irregular intervals chiefly on the disc of the elytra between the striae; the sides arcuately rounded to the apex. Underside coloured as above; the sterna finely and less closely punctured. Legs pitchy; tibiae and tarsi paler, covered with fine dirty yellow pubescence. Length, 7-7½ mm.

Appears to be allied to *Listroderes subcinctus*, Bohem., from Chili, with which it agrees in form and sculpture; it differs, however, in being larger and more convex, in having the prothorax proportionately shorter and the underside less closely punctured.

AMATHYNETES, gen. nov.

General characters of typical *Listroderes* (*L. costirostris*, Gyllh.); rostrum considerably longer than the head, feebly emarginate at the apex. Antennae moderately long; scape nearly reaching the middle of the eye, somewhat sinuous and clavate at the apex; funiculus seven-jointed, second joint considerably shorter than the first, the third to sixth joints small and globose, the seventh joint very slightly longer and much wider; the club rather elongate. Scrobes deeply impressed, lateral, slightly sinuate in front, almost reaching the lower margin of the eye. Prothorax transverse, the sides scarcely produced; ocular lobes very feebly developed. Elytra ovate, regularly convex without any callosities near the apex. Metasternum short. First and second abdominal segments of about equal length, the third and fourth rather shorter, but together much longer than either of the preceding, the fifth segment slightly longer than the first. Legs moderately robust; femora somewhat clavate, tibiae unarmed. (Type, *A. alticola*.)

This genus chiefly differs from *Listroderes* in the form of the antennal scrobes, the structure of the abdominal segments, and the feebly developed ocular lobes.

19. *Amathynetes alticola*, sp. n.

Hab. Chimborazo (12-15,800 feet).

Elongate-ovate, much narrower in front than behind, pitchy black, somewhat shining, moderately densely covered with small elongate dirty yellow scales and fine pubescence. Head rather convex, finely and very closely punctured; rostrum robust, narrower behind than in front, as closely punctured and pubescent as the head, with a feebly elevated longitudinal line extending from just before the eyes to the apex. Antennae pale pitchy red. Prothorax broadly transverse, slightly narrower in front than at the base, closely but rather finely and obsoletely punctured, the disc more finely and sparingly pubescent than the margins; no apparent median line; sides regularly arcuate. Scutellum finely punctured. Elytra about two and a half times as long as the head and prothorax together, moderately strongly striate-punctate, the interstices rather broad and very slightly raised; humeral angles not prominent; sides arcuately rounded to the apex. Underside rather sparingly covered with long dirty yellow pubescence, very sparingly and obsoletely punctured. Legs pitchy and finely pubescent. Length, $6\frac{1}{2}$-$7\frac{1}{2}$ mm.

The scales and pubescence of this and the following species are very easily removed; some of the specimens collected by Mr. Whymper are somewhat rubbed, and thus have a speckled appearance as if the scales, etc., were arranged in small patches.

20. *A. simulans*, sp. n.

Hab. Chimborazo (15,000 feet).

Very closely allied to *A. alticola*; slightly longer and narrower; the head more closely punctured; prothorax somewhat narrower and decidedly longer, finely but more distinctly and closely punctured; scutellum small, covered with fine punctures and pubescence; elytra rather longer and the sides more parallel, the punctures slightly more distinct. Underside rather more densely pubescent; abdominal segments much more strongly and closely punctured. Length, 8 mm.

21. *Macrops colorum*, sp. n.

Hab. Pichincha (15,500 feet); Chimborazo (16,000 feet).

Elongate, pitchy black, somewhat shining, sparingly covered with short black pubescence. Head finely and densely punctured and rugulose, slightly elevated in the middle; eyes rather large, not very prominent; rostrum considerably longer than the head, rather broader in front than behind, very closely and finely punctured, with three small elevations near the apex. Antennæ ferruginous, the club dusky. Prothorax slightly longer than broad, narrowed in front, only slightly convex, very closely and moderately strongly punctured and finely rugulose; the median line indistinct; the anterior margin somewhat thickened; sides rather strongly arcuate. Scutellum very

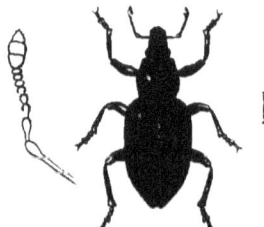

MACROPS COLORUM, OLLIFF.
PICHINCHA, 15,500 FEET; CHIMBORAZO, 16,000 FEET.

small, triangular and impunctate. Elytra more than twice as long as the head and prothorax together, broader behind than in front, rather convex, moderately strongly punctate-striate, the interstices not very broad, slightly raised, very finely rugulose but not punctured; shoulders very slightly prominent; sides gradually arcuate. Underside pitchy black, sparingly covered with short grey pubescence, very finely rugulose; the sterna finely and moderately closely punctured; abdominal segments more finely and less closely punctured. Legs pitchy; tips of the tibiæ and tarsi ferruginous. Length, 5 mm.

Allied to *Macrops humilis*, Gyll. (*M. maculicollis*, Kirby), from North America; differs in having the rostrum broader, slightly shorter and without the longitudinal elevation, the prothorax longer and more narrowed in front, and the underside much less strongly and rather more sparingly punctured.

22. *Anchonus monticola*, sp. n.

Hab. Tortorillas, Chimborazo (12-13,000 feet).

Elongate, rather narrow, somewhat flattened above, dull black. Head very short, extremely finely rugulose; eyes lateral, transverse, placed at the side of the rostrum; rostrum about two-thirds the length of the prothorax, wider in front than behind, rather strongly constricted in the middle, strongly rugose-punctate. Antennae pitchy black, the club covered with grey pubescence. Prothorax longer than broad, moderately strongly constricted in front, strongly rugose-punctate and with small scattered tubercles on the disc, with a strong longitudinal smooth impression extending from just behind the anterior margin to the base; sides rather strongly arcuate. Scutellum absent. Elytra considerably more than twice as long as the head and prothorax together, narrower in front than behind, somewhat constricted just behind

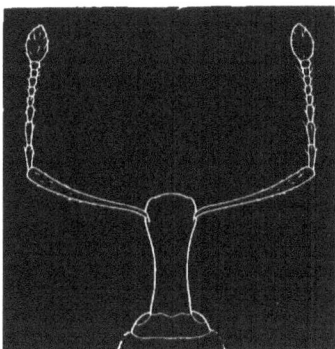

ANCHONUS MONTICOLA, OUTLL.
CHIMBORAZO, 12-14,000 FEET.

the posterior two-thirds, strongly rugose-punctate, sparingly and indistinctly tuberculate on the disc, the sides rather smoother and more evidently punctured, very sparingly covered with short erect hairs; with a short longitudinal elevation at the base on each side of the suture and a much stronger and similarly placed elevation at the apex; the humeral angles raised; sides arcuate. Underside dull black; the sterna somewhat rugose, moderately strongly but obsoletely punctured; the abdominal segments much more strongly punctured. Legs black, tarsi pitchy. Length, 9 mm.

This species has something of the appearance of *Anchonus celsus*, Boheni., but structurally is quite different; the rostrum is narrower, the prothorax shorter and broader, and the sculpture of the upper surface is quite dissimilar as will be seen upon comparing the descriptions.

23. *A. Altarensis*, sp. n.

Hab. Valley of Collanes, Altar (12,500 feet).

Elongate-ovate, rather flattened above, dull black. Head shining and impunctate; rostrum broader in front than behind, strongly constricted in the middle, very strongly rugose-punctate, the punctures arranged regularly at the base. Antennae pitchy, the club paler and densely clothed with grey pubescence. Prothorax rather longer than broad, slightly constricted in front, very strongly rugose; with two slight longitudinal elevations one on either side of the middle; the sides rather strongly arcuate. Scutellum absent. Elytra ovate, somewhat broader behind than in front, rugose and obsoletely punctured, rather densely clothed with short erect hairs, with irregular rows of moderately strong tubercles on the disc; these tubercles disappear towards the sides and the punctures become more evident; shoulders slightly elevated, the sides arcuately rounded to the apex. Underside black; the pro- and mesosternum finely rugulose and punctured in the middle, more strongly rugulose at the sides; the metasternum strongly punctured; the first, second and terminal segments of the abdomen as strongly and rather more closely punctured than the metasternum; the third and fourth segments with a single transverse row of moderately strong punctures. Legs ferruginous, the tibiae pitchy black. Length, 8 mm.

This insect is very distinct from all the described species known to me, but appears to be closely allied to if not conspecific with *Anchonus dorsiger* of Jekel's MS. It differs from *A. monticola* in being broader, in having the rostrum and prothorax more strongly rugose, the elytra more strongly tuberculate, the underside (especially the abdominal segments) more closely punctured, and the legs ferruginous.

24. *Hilipus mysticus*, Pascoe, Trans. Ent. Soc. Lond., 1881, p. 67, pl. i, fig. 5.

Hab. Milligalli (6230 feet).

Var. *nigripes.* Legs black.

A single example which differs from the typical form, described from Sarayacu, in having the femora black instead of flavous. For this well-marked variety I propose the name *nigripes*.

25. *H. longicollis*, sp. n.

Hab. Hacienda of Antisana (13,300 feet).

Elongate-ovate, strongly convex, dull black. Head somewhat polished, narrowed in front, obsoletely and rather sparingly punctured, with a small and moderately strong impression between the eyes; eyes lateral, ovate, not very prominent; rostrum about as long as the prothorax, moderately strongly curved, much narrower in the middle than at the base or apex. Antennæ pitchy black, the club covered with grey pubescence. Prothorax rather long, considerably narrowed in front, slightly flattened, very finely and sparingly punctured on the disc, strongly rugulose at the sides; the median line not distinct but rather strongly impressed at the base and just beyond the middle; with two obsolete impressions on each side, one placed obliquely near the apex and the other behind the middle nearer the side; the anterior margin slightly

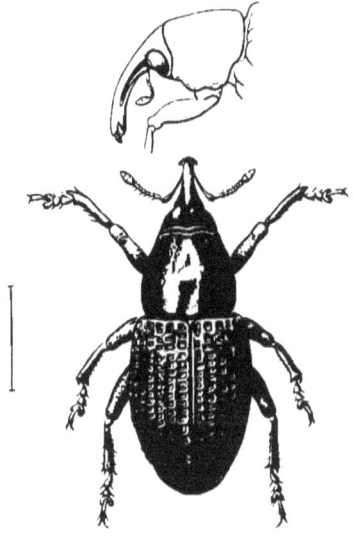

HILIPUS LONGICOLLIS, OLLIFF.
HACIENDA OF ANTISANA, 13,300 FEET.

emarginate in the middle; sides strongly arcuate; posterior margin nearly straight. Scutellum very small, rounded behind and impunctate. Elytra oblong-ovate, convex, about twice as long as the prothorax; sides gradually arcuate; each elytron with seven rather broad depressed costæ which are irregular and more or less interrupted, the interstices about the same width as the costæ, each with a series of flattened tubercles placed rather irregularly. Underside black, shining; the sterna and first four abdominal segments very strongly and sparingly punctured. Legs pitchy black, moderately strongly punctured. Length, 14 mm.

Nearer to *Hilipus scabripennis*, Bohem., from Brazil, than to any other species known to me.

26. *Erirrhinoides distinctus*, sp. n.

Hab. Chimborazo (15,800 feet).

Oblong-ovate, broader behind than in front, black and shining. Head slightly narrower in front than behind; rather strongly and closely punctured, the punctures slightly closer near the eyes, with a feeble impunctate impression near the base of the rostrum; eyes lateral, not very prominent; rostrum moderately long, slightly curved, irregularly punctured near the base, the punctures gradually becoming obsolete towards the apex. Antennae pitchy, the club covered with grey pubescence. Prothorax slightly broader than long, strongly constricted in front, rather strongly and sparingly punctured; the median line slightly raised but not very distinct; the sides strongly arcuate behind the constriction. Scutellum small, triangular and impunctate. Elytra about as long as the rostrum, head, and prothorax together, slightly broader behind than in front, moderately strongly punctate-striate, the interstices rather broad, slightly raised and finely rugulose but not punctured; the striae disappear posteriorly; humeral angles not prominent; the sides gently rounded to the apex. Underside pitchy, the sternal channel extending to the middle coxae; the sterna strongly and not very closely punctured; the abdomen with an oblong impression in the middle common to the first two segments which are as strongly punctured as the sterna, the other segments less strongly and closely punctured. Legs pitchy. Length, $4\frac{1}{2}$ mm.

A very distinct species, differing from the only other species of the genus, *Erirrhinoides unicolor*, Blanch., from Chili, in having the prothorax more strongly constricted in front, the sternal channel broader and extending further posteriorly, etc.

27. *Erirrhinus glaber*, sp. n.

Hab. Cayambe (15,000 feet).

Elongate-ovate, much narrower in front than behind, moderately convex, shining black. Head very short and finely punctured; rostrum long, slightly narrowed towards the base, extremely finely punctured. Antennae pale pitchy, densely clothed with fine grey pubescence. Prothorax slightly broader than long, much narrower in front than behind, rather densely covered with fine indistinct punctures; the median line slightly raised; anterior margin with a slight emargination in the middle; the sides feebly arcuate, with two small obtuse processes one just before, the other just behind the middle. Scutellum small, rounded behind, finely aciculate. Elytra about as long as the rostrum, head and prothorax together, narrower in front than behind, moderately strongly striate-punctate, the interstices rather broad, slightly raised and smooth; shoulders not very prominent; sides arcuately rounded to the apex. Underside black; the sternum rather strongly and closely punctured; abdominal segments shining extremely finely and not very closely punctured. Legs pitchy. Length (including rostrum), 7 mm.

An isolated species which appears to belong to *Ericrhinus*. In general appearance it is nearer to *Erycus* (*Ericrhinus*) *morio*, Mannerh., from North America than to any other species with which I am acquainted.

28. *Ericrhinus*, sp.

Hab. Eastern slopes of Corazon (12,000 feet).

A single specimen of a small testaceous species having the head and rostrum pitchy, in too bad condition for description.

29. *Otidocephalus? spinicollis*, sp. n.

Hab. Chimborazo (12-13,000 feet).

Elongate-ovate, moderately convex, shining black. Head transverse,

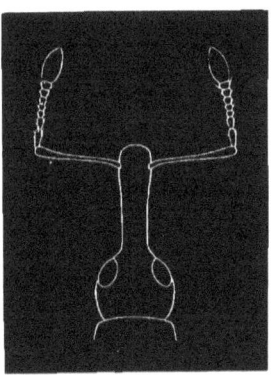

OTIDOCEPHALUS? SPINICOLLIS, OLLIFF.
CHIMBORAZO, 12-13,000 FEET.

finely and not very closely punctured; eyes large, lateral, not prominent; rostrum moderately long, broader in front than behind, finely and rather closely punctured. Antennæ pitchy red, club slightly darker and clothed with yellowish grey pubescence. Prothorax only slightly longer than broad, somewhat narrowed both in front and behind, a little flattened on the disc, rather finely and sparingly punctured; sides arcuate, slightly sinuate in front; with a short, sharp and moderately strong spine on each side near the base. Scutellum very small, rounded behind, impunctate. Elytra about as long as the rostrum, head and prothorax together, at the base considerably broader than the prothorax, gradually widening to just behind the middle, strongly striate-punctate, the interstices not very broad and smooth, with a feeble impression on each side near the shoulder; sides arcuately rounded to the apex. Underside black, shining; the sterna and first abdominal segment very strongly and sparingly punctured in irregular transverse rows, the other abdominal segments with only a few extremely fine punctures near the sides. Legs black, tibiæ and tarsi pitchy. Length, 3¾ mm.

As this insect is only represented by a single example I think for the present it will be better to regard it as an aberrant *Otidocephalus*; perhaps, when more specimens are available for examination, it will be found necessary to establish a new genus for its reception. This insect may be readily distinguished from all the described species by its comparatively broad spined prothorax and strongly punctured surface.

30. *Lyterius*, sp.

Hab. Pichincha (12,000 feet).

A single broken example of a species which appears to belong to this genus.

31. *Apion Andinum*, sp. n.

Hab. Panecillo, Quito (10,000 feet).

Short ovate, dull brassy green, shining, convex, rather thickly covered with short grey pubescence. Head rather narrow, finely and irregularly punctured; rostrum about as long as the head and prothorax together, slender, not very strongly curved, not pubescent; eyes rather prominent, moderately large. Prothorax at the base about one-third broader than long, narrowed and slightly constricted anteriorly, rather strongly and not very closely punctured; anterior margin slightly raised, the angles rounded, sides arcuate behind the constriction; posterior angles nearly right angles. Scutellum rounded behind. Elytra not quite twice as long as broad, slightly prolonged behind, rather strongly striate, the striæ obsoletely punctured, the interstices broad, somewhat raised and impunctate; humeral angles not very prominent; sides arcuately rounded, slightly sinuate just before the apex, which is somewhat produced. Underside not such a metallic colour as above, more finely pubescent, obsoletely punctured. Legs moderately robust, finely pubescent. Length, 2¼ mm.

As this species does not appear to agree with any of the South American species of *Apion* described by Dr. Gerstaecker in the "Stettiner Entomologische Zeitung" for the year 1854, I have ventured to characterise it as new. In general shape and colour it is not unlike *A. cuprescens*, Mannerh., from North America, but in sculpture, form of the rostrum, etc., it differs greatly as will be seen upon comparing the descriptions.

Fam. *CALANDRIDÆ*.

32. *Metamasius sericeus*, Latr.   Humboldt and Bonpland, Voy. i, p. 206, pl. 22, fig. 4 (1811); Horn, Proc. Amer. Phil. Soc. xiii, p. 410 (1873).

This widely distributed species was found at Nanegal at an elevation of three thousand feet.

33. *Sphenophorus notandus*, sp. n.

Hab. Milligalli (6230 feet).

Elongate-ovate, dull pitchy black, flattened above. Head small, extremely finely and not very closely punctured; rostrum long, strongly curved, pitchy red, with a moderately strong longitudinal impression extending from the base to just in front of the antennae. Antennae pitchy black and shining. Prothorax long, narrowed in front, very strongly and irregularly punctured; the sides nearly parallel for the basal two-thirds of their length then narrowed to the apex. Elytra not much longer than the prothorax, strongly narrowed towards the apex, with fine striae, the interstices broad and with very feeble transverse elevations near the suture. Underside black; prosternum dull, as strongly punctured as the prothorax; mesosternum shining, extremely finely punctured; metasternum and abdominal segments shining and not quite as strongly punctured as the prosternum. Legs pitchy black; tibiae pitchy red. Length, 15 mm.

At some future time it may be found necessary to institute a genus for the reception of this species; the anterior coxae are quite as widely separated as in *Metamasius*, Horn. In general appearance and colour it greatly resembles *Sphenophorus pustulatus*, Gyll., but differs in having the prothorax less narrowed in front and the sides more parallel.

34. *Calandra setulosa*, Gylh. Schönh., Gen. Curc. iv, p. 969 (1838).

Hab. Bodegas (level of sea).

One specimen of this beautiful little species which does not seem to differ from the ordinary Mexican form. I have recently seen this species in Mr. Fry's collection from the island of Trinidad.

35. *Cossonus coloratus*, sp. n.

Hab. Pichincha (12-13,000 feet).

Elongate, much depressed pitchy black, shining; the middle and sides of the prothorax pitchy red; the elytra, except the suture and lateral margins, pale pitchy yellow. Head finely and closely punctured; rostrum rather long, greatly dilated in front, constricted behind, finely and moderately closely punctured behind the antennae, more finely punctured nearer the apex. Antennae pitchy red, the club somewhat darker and covered with fine grey pubescence. Prothorax somewhat longer than broad, much narrowed and rather strongly constricted in front, the disc flattened, strongly and rather closely punctured in the middle, the punctures finer and much closer near the sides, with a broad and very feebly raised oblique pitchy line on each side of the middle extending from the anterior to the posterior margin; sides arcuate. Scutellum small, pitchy red, impunctate. Elytra more than twice as long as the head and prothorax together, at the base considerably broader than the prothorax, moderately strongly and very closely seriate-punctate, the interstices rather narrow, slightly raised and impunctate; shoulders not very prominent; sides parallel, rounded behind. Underside pitchy black; prosternum strongly and very closely punctured; mesosternum, metasternum and the first abdominal segment strongly but less closely punctured, the other abdominal segments finely and not very closely punctured. Legs pitchy black, the tarsi pitchy red. Length, 5 mm.

A very distinct species quite unlike anything known to me.

Fam. BRENTHIDÆ.

36. *Estenorchinus designatus*, Bohem.(?) Schönh., Gen. Curc. v, p. 466 (1840).

Hab. Milligalli (6230 feet).

A single discoloured specimen, apparently referable to this species, which was described from Colombia.

37. *Breathus vulneratus*, Gylh. Schönh., Gen. Curc. i, p. 345 (1833).

Hab. Milligalli (6230 feet).

APPENDIX—COLEOPTERA.

One very large specimen of this Brazilian species, measuring 42 mm. inclusive of the rostrum, and having the spots on the elytra unusually broad, and brightly coloured. [A figure of this remarkable form is given, as it does not appear to have been engraved before.—*E. W.*]

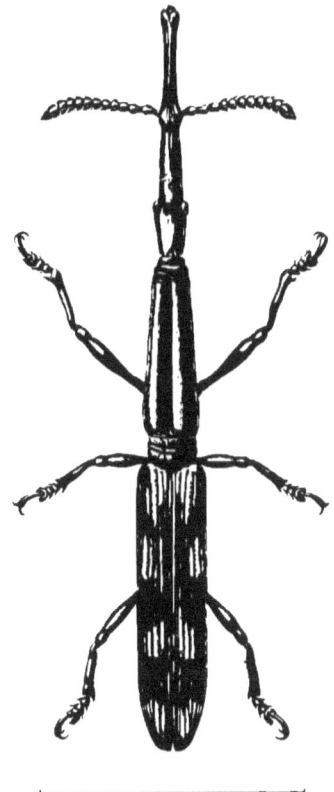

BRENTHUS VULNERATUS, GYLL.
MILLIGALLI, 6230 FEET.

COLEOPTERA (Continued).

By MARTIN JACOBY.

Fam. EUMOLPIDÆ.

1. *Nodu atra*, Harold, Coleopterol., Hefte xiii, 1875, p. 31.

Hab. Pacific slopes (1-3000 feet). One example. Has also been obtained in Colombia and Mexico.

2. *Colaspis callichloris*, Lefèvre, Mitth. Münchener Ent. Verein, 1878, p. 121.

Hab. Nanegal (3-4000 feet). One example. Has also been obtained in Colombia and Mexico.

3. *C. montana*, n. sp.

Hab. Nanegal (3-4000 feet). One example.

Obscure piceous below; femora pale fulvous; joints 5-7 and the 11th of antennae black; above obscure dark fulvous with greenish gloss; thorax finely, elytra strongly semipunctate-striate.

Length, $6\frac{1}{2}$ mm.

Head rather closely and strongly punctured, deeply transversely grooved between the eyes; epistome about as broad as long, punctured like the head; labrum fulvous; jaws black; antennae slender, the third and fourth joints of equal length, the four basal joints and the eighth to the tenth, testaceous,—the others black; thorax rather transverse, scarcely narrowed in front, sides deflexed anteriorly, the angles produced in a short tooth; lateral margin bidentate at the middle, surface finely punctured, dark fulvous, the extreme lateral margins metallic green; elytra distinctly transversely depressed below the base, very strongly punctate-striate anteriorly, geminate and more finely punctate below the middle, the punctuation very coarse and deep at the sides where the interstices are convex and transversely wrinkled, from the shoulder to the apex runs a more or less distinct costa parallel with the lateral margin, another short costa is placed close to the first one at the humeral callus; underside, the knees, tibiae and tarsi piceous; femora fulvous.

The only specimen before me seems closely allied in coloration to *C. luridula*, Lefèvre, but differs in its larger size, more finely and closely punctured thorax and in the colour of the intermediate and ultimate joints of the antennae as well as the tibiae and tarsi.

4. *Alethaxius* (*Alætes*) *nigritarsis*, n. sp.

Hab. Forests above the Bridge of Chimbo (1-3000 feet). One example.

Ovate, obscure piceous below; terminal joints of antennæ, apex of tibiæ and the tarsi, black; above obscure æneous; thorax very finely punctured; elytra more strongly punctate-striate, the apex costate.

Length, 4¼ mm.

Head strigose-punctate at the vertex, with an obsolete longitudinal central groove, obsoletely transversely depressed between the eyes and rugose-punctate at the same place; epistome more strongly punctured than the head; labrum fulvous; jaws black; antennæ more than half the length of the body, slender, third joint double the length of the second, four or five basal joints fulvous, the rest piceous; thorax nearly three times as broad as long, slightly widened at the middle and convex at the same place, all the angles acute and produced in a pointed tooth; lateral margins distinctly bisinuate; surface extremely finely and closely punctured; elytra scarcely wider at the base than the thorax, with a transverse depression below the base, very regularly punctate-striate, the punctures rather distantly placed, and getting gradually deeper towards the sides; interstices strongly costate near the apex, transversely rugose below the shoulders, where a short longitudinal costa is also visible; legs fulvous, apex of the tibiæ and the tarsi, black. A single apparently female specimen.

Principally separated from several closely allied species described by M. Lefèvre, by the closely and finely punctured thorax, the bisinuate sides of the latter and the general smaller size of the insect as well as its black tarsi.

Fam. CHRYSOMELINÆ.

5. *Calligrapha nupta*, Stål, Monogr. Chrysom., 1862, p. 267.

Hab. Nanegal (3-4000 feet). One example. Has also been obtained in Colombia.

6. *C. Percheronii*, Guér., Voy. Coquille Zool., 1830, p. 146. Stål, Monogr., p. 272.

Hab. Riobamba (9000 feet), Hacienda of Guachala (9217 feet). Nine examples. Has also been obtained in Peru.

7. *Doryphora funebris*, var. Jacoby. Proc. Zool. Soc., 1880, p. 597. Tab. liv, Fig. 11.

Hab. Nanegal (3-4000 feet). Previously obtained by Mr. Buckley in Ecuador.

The specimen obtained by Mr. Whymper is no doubt but a variety of the above in which the isolated greenish aeneous spots of the typical form are united in shape of two very broad transverse bands, occupying almost the entire disk and interrupted only by the very narrow flavous bands; the shape of the latter agrees quite with the yellow markings of the type, but the punctuation is rather more closely placed; the insect is also larger (probably a female). In other respects no differences are to be found.

8. *D. picturata*, n. sp.

Hab. Tanti (1890 feet). One specimen.

Ovate, convex; below greenish black; head and lateral margins of the thorax, flavous; elytra bluish black, geminate punctate-striate, a round spot near the scutellum, a narrow short band near the lateral margin, a broader transverse one below the middle and a spot near the apex, flavous.

Length, 7¼ mm.

Head rather closely and finely punctured with several obsolete depressions; antennae extending below the base of the thorax, the first five joints metallic greenish black, the rest opaque, pubescent; thorax nearly three times as broad as long, the sides straight near the base, rounded towards the apex where they turn inwards towards the head; anterior angles pointed; surface sparingly and rather finely punctured, middle of the disk, greenish black, the sides broadly flavous; elytra not broader at the base than the thorax, strongly and closely geminate punctate-striate, greenish black, a narrow short longitudinal band at the base, in front of the lateral margin, a round spot near the scutellum, a broad transverse, slightly curved band below the middle, not touching the margins and a small triangular spot at the apex, flavous; mesosternal process very short. For Figure see the accompanying Plate.

9. *Chrysomela cisseis*, Stål, Monogr. Chrys. Amer., p. 335.

Hab. Nanegal (3-4000 feet). One example. Has also been obtained in Peru.

Fam. *HALTICIDÆ*.

10. *Epitrix nigroaenea*, Harold, Coleopt., Hefte xiv, 1875, p. 36.

Hab. Hacienda of Guachala (9217 feet), Quito (9500 feet), Village of Cayambe (9320 feet). Six examples. Has also been obtained in Colombia.

11. *Longitarsus oopterus*, Harold, Coleopt., Hefte xv, 1876, p. 29.

Hab. La Dormida, Cayambe (11,800 feet). Has also been obtained at La Luzera, Colombia.

The only specimen obtained, agrees in every respect with the description of the author; it is wingless and of an uniform chestnut brown colour: for the rest I must refer to the original description.

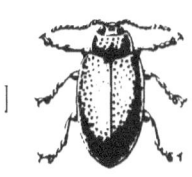 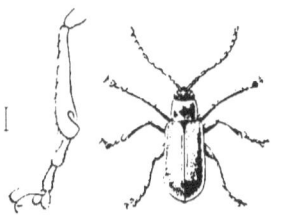

DIBOLIA VIRIDIS, JACOBY.   LUPEROSOMA MARGINALIS, JACOBY
EASTERN SIDE OF CORAZON, 12,000 FEET.   PANECILLO QUITO, 10,000 FEET.

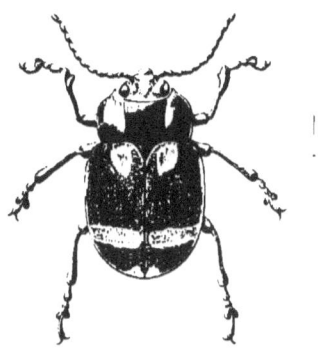

DORYPHORA PICTURATA, JACOBY.
FROM TANTI, 1890 FEET.

12. *Haltica amethystina*, Olivier, Ent., vi, p. 687, Tab. 2, Fig. 31.

Hab. Machachi (9-10,000 feet) and Chillo (9000 feet). Fourteen examples. Has previously been obtained in S. Domingo, Guatemala, Honduras, and Colombia.

13. *Trichaltica costatipennis*, n. sp.

Hab. Hacienda of Guachala (9217 feet). One example.

Underside and legs, black; head and thorax yellowish brown, coarsely punctured; antennae black, apical joints flavous; elytra blue-black, deeply punctate-striate, the interstices distinctly costate and finely pubescent.

Length, 2 mm.

Head very deeply punctured round the inner margin of the eyes; the frontal tubercles indicated only by very narrow oblique ridges; antennae about half the length of the body, black, the three lower joints flavous, spotted with black above; second joint thicker but scarcely shorter than the third; thorax about one-half broader than long, slightly narrowed at the base, the sides widened at the middle, anterior and posterior margin parallel, surface flavous like the head, transversely sulcate at the base, very coarsely but remotely punctured, the sulcation impressed with a single row of closely placed punctures; scutellum black; elytra parallel, of a very dark bluish black, each elytron with about ten rows of regular and very deep transversely shaped punctures, the interstices closely longitudinally costate and sparingly covered with whitish hairs; underside and legs black.

At once separated from the species described by von Harold and Mr. Baly by the costate elytra, their transversely shaped punctuation and colour, as well as that of the underside and legs. A single specimen was obtained.

14. *Aphthona Ecuadoriensis*, n. sp.

Hab. Eastern side of Corazon (12,000 feet). One specimen.

Obscure greenish black below; legs piceous; three basal joints of antennae, fulvous; above dark bluish black; head and thorax finely punctured; elytra deeply and rather regularly punctate-striate.

Length, 3 mm.

Head rather closely and finely punctured; frontal tubercles in shape of narrow transverse elevations; carina distinct and acutely raised, palpi slender, filiform; antennae less than half the length of the body, black, three lower joints fulvous, third and fourth joints equal, one-half longer than the second; thorax subquadrate, a little wider than long, the angles distinct but not produced, surface rather closely and finely punctured, greenish black, shining; elytra bluish, deeply and rather regularly punctate-striate, the striae closely arranged; posterior first tarsal joint slightly longer than the two following, united; claws appendiculate; posterior femora very moderately thickened.

The strongly and regularly punctate-striate elytra distinguish this species principally from other South American forms; I am somewhat doubtful if the true place of the present insect is in *Aphthona*.

15. *Dibolia viridis*, n. sp.

Hab. Eastern side of Corazon (12,000 feet). Three examples.

Ovate, testaceous; terminal joints of the antennae, black; above light green, thorax transverse, remotely punctured; elytra closely and strongly semistriate-punctate, the interstices slightly rugose.

Length, 3½ mm.

Head obscure testaceous, impunctate, transversely grooved between the eyes; frontal tubercles oblique, distinct; clypeus swollen, triangular; antennae less than half the length of the body, the second and third joints equal, short, fourth slightly longer, the rest gradually widened; basal and third joint testaceous, the second and terminal joints black; thorax nearly three times as broad as long, narrowed in front, the sides nearly straight, posterior margin rather strongly rounded; anterior angles obtuse and thickened; surface distinctly but very remotely punctured at the disk, more closely at the sides with an obsolete fovea near the posterior angles; scutellum obscure testaceous; elytra narrowed towards the apex, not wider at the base than the thorax, with a longitudinal depression within the humeral callus; surface very closely and rather strongly punctured, the punctuation arranged in irregular rows near the suture, the interstices, especially at the sides, slightly transversely wrinkled, the space near the lateral margin, impunctate; legs robust, posterior femora strongly incrassate, their tibiae deeply channelled and dilated at the apex, the latter armed with a distinct bidentate spur; first posterior tarsal joint scarcely as long as the two following united; claws appendiculate; prosternum broad, pubescent. For Figure see the Plate facing p. 84.

The species described here, which may be known at once by its pale green colour, seems to be the first of the genus which has been obtained in South America.

16. *Diphaulaca glabrata*, n. sp.

Hab. Eastern side of Pichincha (12-13,000 feet). Two examples.

Oblong-ovate, black below; above metallic green; first three joints of the antennae obscure fulvous; thorax impunctate; elytra extremely finely punctured.

Length, 3-4¼ mm.

Head impunctate; frontal tubercles extremely small, scarcely visible; carina short and thick; antennæ two-thirds the length of the body, black, the three basal joints dark fulvous, third and fourth joints of equal length, one-half longer than the second; thorax one-half broader than long, the sides very nearly straight, all the angles acute, the basal groove deep, sinuate and laterally bounded by a deep longitudinal groove, surface smooth, impunctate; elytra with the base but slightly elevated, extremely finely punctured, the punctuation arranged here and there in longitudinal lines; legs with a slight metallic gloss; abdomen black.

Several closely allied species have been described from Colombia from all of which I must separate the present insect by the almost complete want of the frontal tubercles, the straight sides of the thorax in connection with the very finely punctured elytra.

Fam. GALERUCIDÆ.

LUPEROSOMA, n. gen.

♂. Body elongate; eyes entire; third joint of palpi, robust, swollen; frontal tubercles very swollen, transverse; antennæ subfiliform, incrassate towards the apex, second and third joints very short, subequal, fourth as long as the two preceding joints together; thorax square-shaped, transversely depressed at the disk; apex of the scutellum obtuse; elytra irregularly punctured, their epipleuræ indistinct below the middle; tibiæ without spine, the intermediate emarginate at the apex, the inner margin produced in shape of a spine; posterior first tarsal joint as long as the two following ones united; claws appendiculate; anterior coxal cavities closed. ♀. intermediate tibiæ without emargination; prosternum not visible.

Type, Luperosoma marginata.

I am obliged to erect this genus for the reception of a small species of *Galeruca* having the appearance of *Diabrotica* or *Luperus*, and distinguished from either and other genera by the unarmed tibiæ, short second and third joints of antennæ and the other characters given above. The curious structure of the tibiæ in the male is another peculiarity of the genus which would enter Chapuis' 26th group, the *Platyxanthinæ*.

17. *Luperosoma marginata*, n. sp.

Hab. The Panecillo, Quito (10,000 feet). Three specimens.

Below black; first three joints of the antennæ and the legs obscure testaceous; above testaceous, the disk of the thorax and a broad longitudinal elytral band piceous.

Var. elytral band very obsolete.
Length, 3 mm.

♂. Head black, impunctate, the frontal tubercles contiguous and forming a highly raised transverse ridge; clypeus triangular, strongly raised; antennæ more than half the length of the body, black, three lower joints testaceous below; the fourth joint the longest, terminal joints gradually widened; thorax scarcely broader than long, narrowly margined, with an obsolete transversely oblique depression at each side, testaceous, the disk more or less piceous, impunctate; scutellum black; elytra very slightly widened towards the apex, extremely finely and irregularly punctured, the apex nearly impunctate; legs obscure fulvous, stained slightly with piceous. For Figure see the Plate facing p. 84.

In one specimen the black longitudinal band of each elytron occupies nearly the entire disk, leaving only the margins light testaceous; in another this latter colour prevails, while the band is only indicated by obscure piceous; the clypeus in the female insect is also testaceous and the intermediate tibiæ are of normal structure, the same parts in the male having the apical portion excavated, so that the inner margin is interrupted and produced in a spine.

18. *Diabrotica erythrodera*, Baly, Annals of Natur. Hist., 1878.

Hab. Milligalli (6230 feet). Has also been obtained in Peru.

The only specimen obtained, differs from the description of Mr. Baly in having the fourth joint of the antennæ *longer* not equal to the two preceding ones united, and in the colour of the elytra which are shading gradually at their posterior portion into brownish, and having also an indistinct ring-shaped mark of the same colour near the apex. In other respects there is no difference to be found.

HYMENOPTERA.

FORMICIDÆ.

By PETER CAMERON.

CAMPONOTUS, Mayr.

1. *Camponotus sylvaticus*, Oliv., Encycl. Méth., vi, p. 491.

Hab. Penipe to Riobamba (9000 feet). Several ♀ minor, Hacienda of Guachala (9217 feet). Seven examples.

The body is deep black; the mandibles piceous to ferruginous; antennæ dark red; the scape darker, and the legs pale reddish. There is very little hair on any part of the body, and the abdominal segments at their junction are white. The species is very generally distributed over the old world, as well as in America.

2. *C. atriceps*, Smith, Cat. of Hymen., vi, p. 44, No. 147.

Hab. Guayaquil (indoors). Numerous males.

Some specimens have the thorax pallid rufo-testaceous, the head for the greater part black, the legs and antennæ coloured like the thorax, except that the scape is darker and the coxæ and trochanters paler; the abdomen at the base is dark testaceous, the apex fuscous. Other specimens have the body and legs fuscous or fuscous-black, or dark brown with the flagellum and tarsi pallid testaceous. Most of the specimens have the abdominal segments whitish at their junction. The texture of the body does not differ from that of the ♀ or ☿, but the pilosity is less, especially on the thorax, which may want it entirely. The wings are sometimes tinted with yellow or pale fuscous anteriorly; the nervures pallid testaceous, pale brown or whitish, and the stigma may be fuscous.

3. *C. Mayri*, sp. n.

Hab. Bodegas (level of sea). Two specimens.

Black, opaque; scape and first joint of flagellum pallid red; mandibles red, black at apex; apical tarsal joints and trochanters piceous. Mandibles with moderately large shallow punctures, with five teeth. Head minutely and closely punctured, the punctuation stronger below the antennæ. Thorax closely punctured all over, stronger than on the head; base of abdomen closely and minutely punctured, the rest of it shagreened. Coxæ punctured. Scape of antennæ covered with a short, more or less erect, white pile; flagellum with a depressed almost microscopic pubescence. Head covered with a white, glistening, moderately long pile, longest on the face, and on the top are a few long, erect hairs. Mandibles marked with a few depressed hairs. Pro-meso-and metanotum covered sparsely with long, erect hairs, and with a sparse pubescence; the pubescence silvery white, the hairs fuscous, longest on metanotum. Pleuræ almost glabrous. Scale of abdomen with a few long, pale hairs. Base of abdomen almost without pubescence; the rest covered closely with a thick, depressed cinereous, intermixed with a

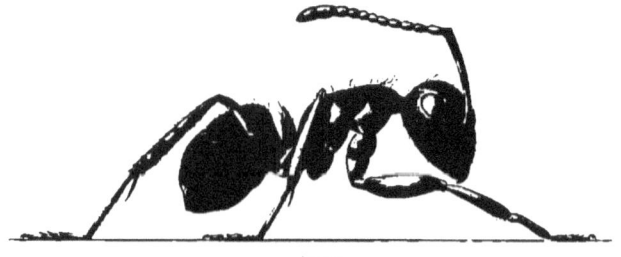

CAMPONOTUS MAYRI, CAMERON.
BODEGAS, LEVEL OF SEA.

few long, white, glistening hairs; the ventral surface covered sparsely with longish, scattered white hairs, which are longest at the base of abdomen. Edges of abdominal segments white. Legs with an erect white pile, thinner and longer on femora. Clypeus and front with an indistinct carina.

Length, 5-6 mm.

In the table given by Mayr in his paper on the ants of Colombia (Sitzb. d. K. Akad. d. Wissens., 1870), of the species of *Camponotus* from that region, the present species would come in at 19, "abdomen with a silky, shining, depressed pubescence," and may be distinguished from the species with this peculiarity, thus:—

> The pubescence on head, thorax, and abdomen yellow. Antennæ and body mostly reddish *auricomus*, Rog.
> The pubescence on head and thorax silvery white, on abdomen cinereous; antennæ and body black, except the flagellum. . . . *Mayri*, sp. n.
> In size, punctuation, and structurally, it agrees closely with *C. crassus*,

Mayr, which has, however, only a sparse yellowish white pubescence on the abdomen, and the flagellum is red.

ECTATOMMA, Smith.

4. *Ectatomma quadridens*, Fab., Ent. Syst., ii, p. 362 = *brunneum*, Smith, Cat. of Hymen., vi, p. 103, No. 2.

Hab. Guayaquil. Five examples.

PACHYCONDYLA, Smith.

5. *Pachycondyla villosa*, Fab., Syst. Piez., p. 409, = *pedunculata*, Sm., Cat. of Hymen., vi, p. 96.

Hab. Bodegas (level of sea). One example.

6. *P. harpax*, Fab., Syst. Piez., p. 401 = *Montezumia*, Sm., Cat. of Hymen., vi, p. 108.

Hab. Pacific slopes (1-2000 feet). Two examples.

7. *P. carbonaria* (*Ponera carbonaria*, Smith, Cat. of Hymen. Form., p. 97 ?).

Hab. Ibarra (7300 feet). Many examples from the garden of Señor Teodoro Gomez de la Torre.

I am not quite certain if the specimens collected by Mr. Whymper are identical with *Ponera carbonaria*, Sm., from Quito. Smith describes his species as "jet-black," while the present species is bluish black, the bluish tinge being very conspicuous, and is present even in the legs, although not so strongly as on the body. The mandibles are deep black, very finely striated ; along the inner edge is a row of large punctures, and it bears also some long, reddish hairs, the outer edge having somewhat shorter pale hairs. In *carbonaria*, Smith, the mandibles are said to be "obscurely ferruginous." The antennæ are black, the apical joints obscurely punctured. The eyes are situated opposite the base of the frontal laminæ, which are finely punctured. The frontal suture is deep ; it originates about the centre of the frontal laminæ, and anteriorly curves round the triangular frontal area, which is not defined from the clypeus. The frontal laminæ are curved, narrower at base than at apex, and from behind the antennæ project into a tubercle-like dilatation. The antennal and clypeal foveæ are united. Clypeus incised at the apex. The front of the head is finely longitudinally striated. Smith describes the pubescence on his *carbonaria* as "yellowish" ; in the present species it is whitish, ferruginous on the metathorax behind, and on the tibiæ and tarsi.

Length, 11-12 mm.

HOLCOPONERA, gen. nov.

Head quadrangular, longer than broad. Eyes oval, small, situated a little behind the middle of the head. Ocelli present. Antennae 12-jointed. First joint of flagellum globose, longer than second. Mandibles triangular, without lateral teeth, apical tooth scarcely separated from the inner edge. Clypeus incised broadly at the apex, concave, not separated from the rest of head. Frontal laminae dilated, extending backwards to the middle of the eyes, rounded in front, converging towards the eyes behind; laterally forming a receptacle into which the entire scape can be retracted. Frontal fovea large, deep, oval behind, open in front, extending into the frontal area which is not defined from the clypeus. Meso- and metanotum without a suture. Petiole longer than broad, contracted somewhat at the sides, concave above, separated from the 1st abdominal segment by a belt-like constriction; carinated beneath in the middle, a blunt tooth at the base. A belt-like constriction between the 1st and 2d abdominal segments. Claws simple, dilated at the base. Head, thorax, petiole and 1st abdominal segment deeply longitudinally sulcated.

The most remarkable peculiarities of this genus are the great development of the frontal laminae, which are greatly dilated, and the deeply sulcated head, thorax and base of abdomen. There are only 9 or 10 of these furrows on the mesonotum, and 7 on the petiole above. Otherwise its affinities appear to be with *Pachycondyla*, but that differs from it in the toothed mandibles, in the form of the thorax, etc. In Mayr's table of genera (Verh. z. b. Wien, 1862, p. 713) it comes in between 9 and 10.

8. *Holcoponera Whympcri*, sp. n.

Hab. Guayaquil (indoors). A single specimen.

Black: mandibles and more or less of clypeus, base and 2d joint of antennae, piceous; apical joint of antennae fulvous; tibiae (except at base and apex) white; four apical joints of tarsi and extreme apex of abdomen ferruginous. Mandibles shining, longitudinally striated, the biting edge smooth. Frontal fovea finely striated; head in front of eyes striated; scape finely striated; flagellum becoming gradually thicker towards the apex, sparsely covered with a depressed pubescence, last joint conical at apex, double the length of preceding. Thorax compressed laterally, the middle contracted. Above the 1st pair of legs is a fine semi-perpendicular suture; above the 2d a wider, more oblique and curved one; behind the first of these sutures the pleurae are longitudinally striated. The metanotum behind is smooth, aciculated, with a slight slope, united to the petiole in the middle;

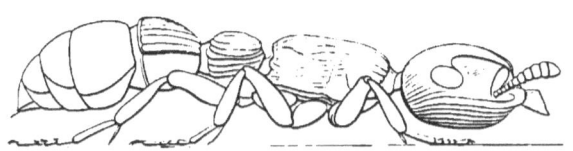
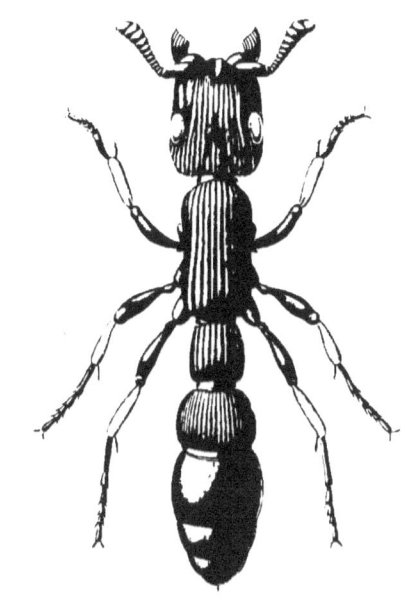

TAKEN INDOORS AT GUAYAQUIL.

above there is a distinct margin separating the posterior aciculated portion from the anterior sulcated region. In front of the petiole, close to its junction with the metathorax, is a blunt tubercle, behind this the petiole is finely striated laterally. Directly in front of the metathorax it is aciculated, above there is a distinct margin. The sides of the petiole are slightly compressed. Separating the petiole from the abdomen is a broad (comparatively) belt-like constriction, smooth, shining, very finely transversely striated. A similar constriction is between the 1st and 2d segments; it is contracted a little at the side, expanding again on the ventral surface. The lower half of the sides of the 1st segment is smooth, shining, impunctate, as are also the remaining abdominal segments; these bear above some short, glistening hairs; beneath some longer pale hairs. Legs stout, smooth shining. For Figure see the accompanying Plate.

Length, 8 mm.

ODONTOMACHUS, Latr.

9. *Odontomachus hæmatodes*, Lin., Syst. Nat., ii, p. 965.

Hab. Guayaquil. Two examples.

PHEIDOLE, Westw.

10. *Pheidole monticola*, sp. n.

Hab. Cayambe village (9320 feet), workers; Penipe to Riobamba (9000 feet), soldiers; the Panecillo, Quito (10,000 feet), females and workers. Numerous specimens.

Soldier. Blackish fuscous; the abdomen more or less obscured with testaceous, thorax with piceous; mandibles ferruginous; more or less of face below the antennæ and basal half of scape obscure ferruginous; apex of last joint of antennæ castaneous. Mandibles shining, inner edge black, acute, inner half with some minute punctures; covered with a pale pubescence longest on lower side; apex bidentate. Head covered with a depressed pile, shining, smooth, obscurely alutaceous in front; frontal laminæ curved, dilated in middle; frontal fovea absent, area raised, a depression at its base; separated clearly from clypeus, which is obscurely punctured, covered with a pale, scattered, depressed pile. Pronotum almost shining, scarcely punctured; meso- and metanotum minutely punctured, semi-opaque; metathorax with two stout, slightly diverging spines; the space between these and behind smooth, shining, almost impunctate. Petiole half-shining, completely so at base, which is testaceous; first node depressed in centre of top; second double wide as long, aciculated, sides projecting, rounded. Abdomen half-shining, obscurely aciculated, the junction of segments pale; covered sparsely with a moderately long, pale hair. Legs covered with a white, glistening, semi-depressed pubescence; the apex of coxæ and trochanters obscure ferru-

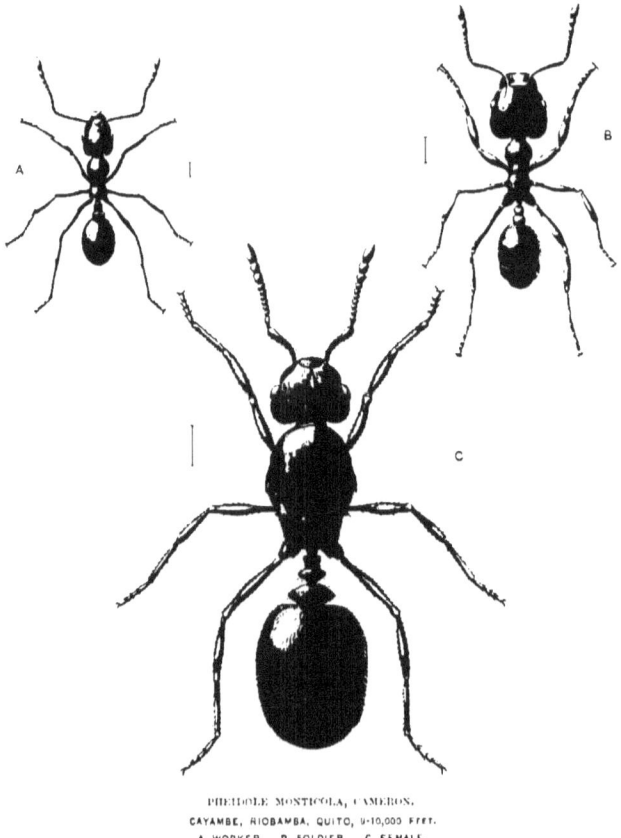

PHEIDOLE MONTICOLA, CAMERON.
CAYAMBE, RIOBAMBA, QUITO, 9-10,000 FEET.
A. WORKER.   B. SOLDIER.   C. FEMALE.

ginous; tibiæ obscure ferruginous or piceous; tarsi testaceous. The thorax bears some depressed, pale hairs; the petiole has a longer pubescence.

Length, 5½ mm.

Worker. Testaceous, obscured with fuscous or black. Head and mandibles smooth, shining, impunctate, front sometimes faintly aciculated; frontal laminæ somewhat shorter than in soldier; sides of pleuræ longitudinally striated and punctured. Posterior node of petiole not so much rounded and dilated at the sides, and not contracted posterior as in soldier; anterior not depressed in centre above. Abdomen aciculated at the base. Otherwise as in the soldier.

Length, 3½-4 mm.

Female. Black; mandibles and antennal tubercles ferruginous; knees, and 4 anterior tibiæ piceous, tarsi testaceous; apex of last and basal joint of flagellum obscure ferruginous. Entire body densely covered with a greyish or fuscous hair. Head at top smooth, obscurely aciculated, and bearing a few punctures; front longitudinally rugose: smooth in the centre: clypeus shining. Mandibles black on inner border, shining, a few scattered punctures on inner side; apex bidentate; thorax half-shining, alutaceous, metanotum aciculated or punctured, sides finely punctured, a short, obscure keel on mesonotum in front. Between the metathoracic spines (which are triangular, acute at top, and diverging) transversely striated; metapleuræ below longitudinally striated; a carina on either side arising from the spines. Petiole with a distinct neck at base, in front of 1st node, which is obscurely transversely striated below, aciculate above, depressed in centre of top; 2nd node broader than long, bulging out in the centre; the griseous hair is especially thick on petiole. Abdomen densely pilose, alutaceous, segmental divisions white. Antennæ and legs covered with a white, stiff hair.

Length, 7-8 mm.

The worker varies considerably in coloration from light testaceous to fuscous.

In the table of the American species of *Pheidole* given by Mayr (Verh. z. b. Wien, 1870, p. 981) the soldier comes in at 176, near *P. flavens*, Rogers, it having no frontal suture; the worker at p. 984, near *P. pusilla*. In the form of the head it approaches *Aphænogaster*. I am indebted to Prof. Gustav Mayr of Vienna for his opinion on it.

PSEUDOMYRMA, Guér.

11. *Pseudomyrma gracilis*, Fab., Syst. Piez., p. 405.

Hab. Bodegas (level of sea); Pacific slopes (1-2000 feet). Two examples.

ATTA, Fab.

12. *Atta sexdens*, Lin., Syst. Nat., i, p. 964 = *sexdentata*, Latr., Hist. Nat. Fourm., p. 228.

Hab. Pacific slopes (1-8000 feet).

LEPIDOPTERA.

RHOPALOCERA.

By F. DUCANE GODMAN, F.R.S., & OSBERT SALVIN, F.R.S.

The collection of Diurnal Lepidoptera submitted to us by Mr. Whymper contains specimens of 105 species, which were obtained at various elevations from 1000 to 16,000 feet above the sea level. As might be expected, novelties amongst the butterflies were not numerous. As regards the lowlands, the collection represents but a fragment of the Lepidopterous fauna,[1] but as regards the higher mountains Mr. Whymper has probably obtained a majority of the species.

Pieris xanthodice and *Colias alticola* occur at the highest elevations, the former of these has a wide range in the Andes, but the latter is only known as yet from the mountains of Ecuador. At 14,500 feet *Lymanopoda tener* and two species of *Pedaliodes* appear, and a little lower, at 14,000 feet, *Lycæna koa* occurs, the latter being also found in Peru. At 13,000 feet we find *Colias dimera*, and at 12,000 feet an undetermined species of *Pieris*; a form of *Acræa thalia* occurs at 11,000 feet, and at 10,500 a species of *Ancylosypha*—the highest ranging of the *Hesperidæ*. At 10,000 feet and a little below it we get a large accession of species, a *Steroma*, *Lymanopoda luana*, *Agraulis glycera*, *Pyrameis huntera* and *P. caryæ*, *Junonia vallida*, *Lycæna andicola*, *Papilio americus*, *Pieris suadella*, *Colias lesbia*, and *Pamphila phylæus*. Between 9000 and 8500 feet we find three species of *Terias*, *Euptoieta hegesia*, *Pieris eleone*, and *Meganostoma cæsonia*; and at 7300 feet *Pieris elodia*. These represent the upland species of Mr. Whymper's collection, and a glance at the names will show that the bulk of them are Andean forms of genera of wide distribution, and that the evidence of anything like a specialised Alpine butterfly fauna can hardly be said to exist.

Fam. NYMPHALIDÆ.

Sub-fam. DANAINÆ.

1. *Danais plexippus* (Linn.). Godm. and Salv., Biol. Centr. Am., Lep.-Rhop., i, p. 1.

   Hab. Country west of Quito (alt. uncertain). One example.

---

[1] It should, however, be observed that it was not our aim to collect in the lower zones.—E. W.

2. *Ituna lamirus* (Latr.).

Heliconius lamirus, Latr. in Humb. and Bonpl., Obs. Zool., ii, p. 126, t. 41, f. 7, 8.

Ituna lamirus, Godm. and Salv., Biol. Centr. Am., Lep.-Rhop., i, p. 5.

Hab. Nanegal (alt. uncertain). One example.

3. *Mechanitis mantineus*, Hew., Equat. Lep., p. 12 ; Ex. Butt., Heliconiidæ, f. 16.

Hab. Bridge of Chimbo (1000 feet). One example.

A species of western Ecuador, discovered by Buckley, and described by Hewitson in 1869.

4. *Ceratinia antonia* (Hew.).

Ithomia antonia, Hew., Equat. Lep., p. 14; Ex. Butt., Ithomia, t. 29, f. 191.

Hab. Bridge of Chimbo (1000 feet). Two examples.

5. *Ithomia consobrina*, sp. n.

Alis anticis nigris, dimidio basali interne et posticis præter marginem externum ferrugineis, anticarum apicibus maculis septem notatis, duabus costæ proximis maximis sordide albescentibus ; duabus maculis alteris obsoletis ad angulum posticarum apicalem ; subtus ut supra anticarum dimidio apicali ferrugineo tincto.

Exp. 2·5 inches.

Hab. Bridge of Chimbo (1000 feet). One example.

Obs. *I. virginianæ* et *I. adelphinæ* valde affinis sed maculis anticarum apicalibus majoribus et sordide (nec pure) albis ; maculis ad cellulæ finem nullis quoque distinguenda.

We have long possessed specimens of this *Ithomia* in our collection obtained in Ecuador by Buckley and others, but we have hitherto refrained from describing it. The species is closely allied to *I. virginiana* and *I. adelphina*, but with apparently constant distinguishing characters. The three species belong to the section or genus *Hyposcada* of our work on Central American Rhopalocera (Biol. Centr. Am., Lep.-Rhop., p. 35).

6. *Ithomia lilla*, Hew., Ex. Butt., Ithomia, t. 18, f. 108.

Hab. Bridge of Chimbo (1000 feet). One example.

Described from Guayaquil specimens.

7. *Ithomia diasia*, Hew., Ex. Butt., Ithomia, t. 5, f. 28.

Hab. Bridge of Chimbo (1000 feet). One example.

8. *Ithomia andromica*, Hew., Ex. Butt., Ithomia, t. 7, f. 38.
   Hab. Bridge of Chimbo (1000 feet). One example.
   A species of wide range, the type being from Venezuela.

9. *Ithomia padilla*, Hew., Ex. Butt., Ithomia, t. 24, f. 149.
   Hab. Bridge of Chimbo (1000 feet). Three examples.

Sub-fam. SATYRINÆ.

10. *Steroma* sp. ?
    Hab. Machachi (10,000 feet). Two examples.
    These specimens are in bad condition, and we are thus unable to determine them satisfactorily. The species is apparently allied to *S. pronophila*, Felder.

11. *Pedaliodes* sp. ?
    Hab. Bridge of Chimbo (1000 feet). One example. In poor condition. Allied apparently to *P. porina*, Hew.

12. *Pedaliodes* sp. ?
    Hab. Machachi (10,000 feet). Two examples. In poor condition. *P. parrhaeba* appears to be its nearest ally.

13. *Pedaliodes* sp. ?
    Hab. Pichincha (11-12,500 feet); Cotocachi (14,000 feet); Cayambe (13-13,500 feet); Machachi (10,000 feet). Fourteen examples. All in poor condition. *P. manis*, Feld., seems to be nearly allied.

14. *Lymanopoda lecena*, Hew., Journ. Ent., i, p. 156, t. 9, f. 1.
    Hab. Machachi (10,000 feet). Two examples.

15. *Lymanopoda tener*, Hew., Ent. Monthly Mag., vi, p. 98.
    Hab. Cayambe (13-14,500 feet); between Antisanilla and Piñantura 11-12,000 feet); Hacienda of Antisana (13,300 feet); Pichincha (11-12,500 feet); La Dormida, Cayambe (11,805 feet); Machachi (10,000 feet); Altar (13,000 feet). Many examples. Described by Hewitson from specimens collected by Buckley in Ecuador.

16. *Lasiophila zapatoza*, var.
    Pronophila zapatoza, Westw., Gen. Diurn., Lep., p. 358.
    Hab. Bridge of Chimbo (1000 feet). One example.
    Closely allied to the Venezuelan form of *L. zapatoza*, of which it is doubtless a localised race. According to Westwood and Hewitson the range of the species includes the Andes from Venezuela to Bolivia.

17. *Corades cuyo*, Hew., P. Z. S., 1848, p. 117, t. 4.
   Hab. Nanegal (alt. uncertain). One example.

Sub-fam. MORPHINÆ.

18. *Morpho peleides*, Koll., Denkschr. Ak. Wien, i, p. 356; Godm. and Salv., Biol. Centr. Am., Lep.-Rhop., i, p. 119.
   Hab. Nanegal (alt. uncertain); Bridge of Chimbo (1000 feet). Three examples.
   A variable species, ranging northwards into Southern Mexico.

Sub-fam. BRASSOLINÆ.

19. *Opsiphanes tamarindi*, Feld., Godm. and Salv., Biol. Centr. Am., Lep.-Rhop., i, p. 128.
   Hab. Nanegal (alt. uncertain). Two examples.

20. *Opsiphanes quirinus*, Godm. and Salv., Biol. Centr. Am., Lep.-Rhop., i, p. 128.
   Hab. Nanegal (alt. uncertain). Two examples, male and female agreeing with our Central American types.

21. *Caligo oileus* (Feld.).
   Pavonia oileus, Feld., Wien Ent. Mon., v, p. 111; Godm. and Salv., Biol. Centr. Am., Lep.-Rhop., i, p. 132.
   Hab. Nanegal (alt. uncertain). Two examples. These and other Ecuador specimens in our collection have the upper surface of the primaries more uniform in tint, the dark outer border being less evident than in typical *C. oileus*.

22. *Caligo dentina*, Druce, Trans. Ent. Soc., 1874, p. 155.
   Hab. Nanegal (alt. uncertain). One example.

Sub-fam. ACRÆINÆ.

23. *Actinote thalia* (Linn.).
   Papilio thalia, Linn., Mus. Ulr., p. 230.
   Hab. Pichincha (11-12,000 feet); Valley of Chillo (9000 feet). Three specimens which belong to one of the many forms of *A. thalia*.

24. *Actinote ozomene* (Godt.).
   Acræa ozomene, Godt., Enc. Méth., ix, p. 241.
   Hab. Nanegal (alt. uncertain). Two examples.

Sub-fam. HELICONIINÆ.

25. *Heliconius peruvianus*, Feld., Wien Ent. Monatschr., iii, p. 396.
    Hab. Bridge of Chimbo (1000 feet). One example.

26. *Heliconius cyrbia*, Godt., Enc. Méth., ix, p. 203.
    Hab. Bridge of Chimbo (1000 feet). Two examples.
    A very characteristic species of western Ecuador, where alone it has been found.

27. *Heliconius erato* (Linn.). Godm. and Salv., Biol. Centr. Am., Lep.-Rhop., i, p. 160.
    Hab. Country west of Quito. Two examples.
    Both these examples are of the form with the base of the secondaries greenish-blue—the *Papilio doris* of Linnæus.

Sub-fam. NYMPHALINÆ.

28. *Colænis delila* (Fabr.). Godm. and Salv., Biol. Centr. Am., Lep.-Rhop., i, p. 168.
    Hab. Country west of Quito. Two examples.
    These specimens have no dark bar on the primaries, and thus agree with Central American examples rather than with those of countries lying to the eastward, where *C. julia* is found.

29. *Agraulis glycera*, Feld., Wien Ent. Monatschr., v, p. 102.
    Hab. Otovalo (8500 feet); Valley of Chillo (9000 feet); Machachi (9800 feet). Eight examples.
    A species of the Andes of Ecuador, Colombia, and Venezuela.

30. *Agraulis andicola*, Bates, Journ. Ent., ii, p. 187 (note).
    Hab. Nanegal (alt. uncertain). Three examples.
    Discovered in Western Ecuador by Mr. Spruce.

31. *Euptoieta hegesia* (Cram.). Godm. and Salv., Biol. Centr. Am., Lep.-Rhop., i, p. 175.
    Hab. Otovalo (8500 feet). Four examples.
    A species of very wide range.

32. *Eresia clara*, Bates, Godm. and Salv., Biol. Centr. Am., Lep.-Rhop., i, p. 189.
    Hab. Bridge of Chimbo (1000 feet). One example.

33. *Phyciodes flavida* (Hew.), Ex. Butl., Eresia, t. 7, f. 61.
   Hab. Bodegas de Babahoyo (level of sea). One example.
   Peculiar to Western Ecuador.

34. *Phyciodes claudina* (Eschsch.). Kotzeb. Reise, iii, p. 212, t. 8, f. 18 a, b.
   Eresia claudina, Hew., Ex. Butl., Eresia, t. 7, f. 52.
   Hab. Bridge of Chimbo (1000 feet). One example.

35. *Eurema lethe* (Fabr.). Godm. and Salv., Biol. Centr. Am., Lep.-Rhop., i, p. 212.
   Hab. Nanegal (alt. uncertain). Three examples.

36. *Pyrameis huntera* (Fabr.). Godm. and Salv., Biol. Centr. Am., Lep.-Rhop., i, p. 218.
   Hab. Machachi (9800 feet). Two examples.

37. *Pyrameis carye* (Hübn.). Godm. and Salv., Biol. Centr. Am., Lep.-Rhop., i, p. 217.
   Hab. Otovalo (8500 feet); Machachi (9800 feet). Four examples.

38. *Junonia vellida* (Fabr.).
   Papilio vellida, Fabr., Mant. Ins., ii. p. 35; Donovan, Ins. New Holl., t. 25, f. 3.
   Hab. Valley of Chillo (8600 feet); Machachi (9800 feet). Six examples.
   These specimens agree very closely with Australian examples and others from various islands in the Pacific Ocean. The fulvous border of the secondaries is somewhat wider in the Ecuador insects, but the difference is trivial. Besides these specimens, we have in our collection one from Peru, brought from there about fifteen years ago by the botanical collector Pearce; another is from Costa Rica, sent us by Rogers in 1879. *Junonia vellida* therefore seems to be established in South America. In connection with this wide extension of its range, it must be remembered that it was once taken in England and described in 1827 by Stephens as *Cynthia hampstediensis*.

39. *Anartia amalthea* (Linn.).
   Papilio amalthea, Cram., Pap. Ex., t. 309, A, B.
   Anartia amalthea, Doubl. and Hew., Gen. Diurn., Lep., t. 24, f. 5.
   Hab. Bridge of Chimbo (1000 feet). Five examples.

40. *Myscelia cyaniris* (Doubl. and Hew.), Godm. and Salv., Biol. Centr. Am., Lep.-Rhop., i, p. 230.
   Hab. Bridge of Chimbo (1000 feet). One example.

41. *Perisama euriclea*, Doubl. and Hew., Gen. Diurn., Lep., t. 28, f. 5.
Catagramma euriclea, Hew., Ex. Butt., Catagramma, t. 12, f. 90, 91.
Hab. Nanegal (alt. uncertain). One example.

This specimen has rather less red at the base of the secondaries beneath, and these wings are slightly more rounded than in typical Venezuelan examples.

42. *Callicore marchalii* (Guér.). Godm. and Salv., Biol. Centr. Am., Lep.-Rhop., i, p. 256.
Hab. Country west of Quito. One example.

43. *Callicore parira*, Guenée, Mém. Phys. Gén., xxi, p. 388.
Hab. Nanegal (alt. uncertain). Two examples.

These specimens (and we have others) agree fairly with Guenée's description, but we cannot be certain of this identification. The species is allied to the Colombian *C. eucleides*, Latr., but has the cross-band of the wings much wider, that of the secondaries extending almost to the base of the wing.

44. *Callicore nystographa*, Guenée, Mém. Phys. Gén., xxi, p. 387.
Hab. Nanegal (alt. uncertain). Four examples.

The insect we identify with Guenée's description is a local form of *C. eucleides* (Latr.). It has the greenish-blue bands of both wings of nearly equal width instead of the band of the primaries being much wider, as in the Colombian butterfly which we call by Latreille's name.

45. *Didonis biblis* (Fabr.). Godm. and Salv., Biol. Centr. Am., Lep.-Rhop., i, p. 277.
Hab. Bridge of Chimbo (1000 feet). Two examples.

46. *Amphirene epaphus* (Latr.). Godm. and Salv., Biol. Centr. Am., Lep.-Rhop., i, p. 281.
Hab. Nanegal (alt. uncertain). Four examples.

47. *Timetes marcella*, Feld., Wien Ent. Monatschr., v, p. 108 ; Godm. and Salv., Biol. Centr. Am., Lep.-Rhop., i, p. 284.
Hab. Nanegal (alt. uncertain). Three examples.

48. *Timetes berania*, Hew., Ex. Butt., Timetes, t. 1, f. 1 ; Godm. and Salv., Biol. Centr. Am., Lep.-Rhop., i, p. 286.
Hab. Country west of Quito. One example.

49. *Timetes eoresia* (Godt.), Enc. Méth., ix, p. 359.
   Marpesia zerynthia, Hübn., Samml. ex. Schmett.
   Hab. Nanegal (alt. uncertain). Two examples.

50. *Adelpha spruceana* (Bates), Ent. Monthly Mag., i, p. 129.
   Hab. Nanegal (alt. uncertain). Two examples.
   Described by Mr. Bates from a specimen taken by Spruce on the western slope of the Andes of Ecuador.

51. *Adelpha eytherea* (Linn.).   Godm. and Salv., Biol. Centr. Am., Lep.-Rhop., i, p. 303.
   Hab. Country west of Quito. One example.

52. *Aganisthos orion* (Fabr.).   Godm. and Salv., Biol. Centr. Am., Lep.-Rhop., i, p. 324.
   Hab. Nanegal (alt. uncertain). Three examples.

53. *Coea cadmus* (Cram.).   Godm. and Salv., Biol. Centr. Am., Lep.-Rhop., i, p. 326.
   Papilio phereeydes, Cramer.
   Hab. Country west of Quito. Two examples.

54. *Prepona chromus*, Guér., Icon. Règn. Anim., Texte, p. 478.
   Prepona hereules, Doubl. and Hew., Gen. Diurn., Lep., t. 47, f. 1.
   Hab. Mindo (4000 feet); Nanegal (alt. uncertain). Three examples.

55. *Prepona amphitoe* (Godt.).   Godm. and Salv., Biol. Centr. Am., Lep.-Rhop., i, p. 322.
   Hab. Nanegal (alt. uncertain). One example.

56. *Prepona amphimachus* (Fabr.).   Godm. and Salv., Biol. Centr. Am., Lep.-Rhop., i, p. 322.
   Hab. Nanegal (alt. uncertain). One example.

57. *Anaea iphis* (Latr.).
   Nymphalis iphis, Latr., in Humb. and Bonpl., Obs. Zool., ii, p. 80.
   Hab. Country west of Quito. One example.

58. *Anaea amenophis* (Feld.).
   Nymphalis amenophis, Feld., Reise d. Nov., Lep., p. 449.
   Hab. Nanegal (alt. uncertain). One example.

Fam. ERYCINIDÆ.

59. *Mesosemia molina*, Godm. and Salv., Biol. Centr. Am., Lep.-Rhop., i, p. 386.

Hab. Bridge of Chimbo (1000 feet). One example.

60. *Siseme spruce i*, Bates, Journ. Linn. Soc., Zool., ix, p. 384.

Hab. Nanegal (alt. uncertain). Five examples.

61. *Emesis mandana* (Cram.). Godm. and Salv., Biol. Centr. Am., Lep.-Rhop., i, p. 443.

Hab. Country west of Quito. One example.

Fam. LYCÆNIDÆ.

62. *Lycæna koa*, Druce, P. Z. S., 1876, p. 239, t. 18, f. 7.

Hab. Machachi (10,000 feet); Hacienda of Antisana (13,300 feet); Cayambe mountain (13-14,000 feet). Seven examples.

These specimens agree fairly with the types from Peru, though the transverse marks of the primaries beneath are not so clearly shown; but we do not think the Ecuadorean insect is separable on this account. There are several examples in the Hewitson collection of Buckley's collecting.

63. *Lycæna andicola*, sp. n.

Alis supra violaceo-cæruleis, ciliis sordide sericeo-albis; subtus griseo-fuscis maculis obscurioribus albo circumcinctis transfasciatis, posticis fascia albida margini externo plus minusve parallela notata, extra eam ad angulum analem ocellis tribus fulvis nigro pupillatis et argenteo atomatis; anticarum apicibus aliquot acutis. Exp. 1·25 inches.

Hab. Guallabamba (7500 feet); Quito (9400 feet); between Cayambe village and Otovalo (9500 feet); Machachi (9800 feet); Cotocachi (12,000 feet). Ten examples.

Obs. L. cassius et L. marina affinis inter alia anticis magis acutis haud difficile distinguenda.

We have an Ecuadorean specimen of this species and another from Colombia. It is a close ally of *L. marina*, but at the same time readily distinguishable, not only by the more pointed primaries, but also by the smaller size and narrower edging of the rows of transverse spots beneath. There are three spots near the anal angle of the secondaries instead of two as in the allied form, and none appear on the upper surface. In the latter respect it resembles *L. cassius*.

APPENDIX—LEPIDOPTERA.

64. *Lycæna* (?)

  Hab. Pichincha (11-12,500 feet); Cotocachi (12,000 feet).

  Two specimens of a species unknown to us, but which are not in sufficiently perfect condition to determine satisfactorily.

Fam. PAPILIONIDÆ.

Sub-fam. PIERINÆ.

65. *Euterpe leucodrosime*, Kollar, Denk. Ak. Wiss. Wien, Math. Cl. i, p. 358, t. 44, f. 3, 4.

  Hab. Nanegal (alt. uncertain). One example.

66. *Euterpe eritias*, Feld., Wien Ent. Monatschr., iii, p. 327; Reise d. Nov., Lep., p. 158, t. 23, f. 13, 14.

  Hab. Bridge of Chimbo (1000 feet). Two examples.

67. *Euterpe zenobia*, Feld., Reise d. Nov., Lep., p. 146, t. 23, f. 5, 6.

  Hab. Nanegal (alt. uncertain). Two examples.

68. *Hesperocharis marchali* (Guér.), Icon. Règne An. Ins., Texte, p. 468.

  Hab. Nanegal (alt. uncertain). Two examples.

69. *Hesperocharis* sp.?

  Hab. Country west of Quito. One example.

  Unknown to us, but not in a condition to describe.

70. *Leptalis avonia*, Hew., Trans. Ent. Soc., ser. 3, v, p. 563; Ex. Butt., Leptalis, t. 7, f. 46-48.

  Hab. Bridge of Chimbo (1000 feet). One example.

71. *Leptalis jethys*, Boisd., Sp. Gén., i, p. 423.

  Hab. Nanegal (alt. uncertain). One example.

  This specimen agrees fairly with some examples of this species which, so far as we know at present, is subject to great variation as regards the black markings of the primaries. In this respect it is intermediate between the extreme forms.

72. *Leptalis nemesis* (Latr.).

  Pieris nemesis, Latr., in Humb. and Bonpl., Obs. Zool., ii, p. 78, t. 35, f. 7, 8.

  Hab. Valley of Chillo (8600 feet). One example.

73. *Terias gaugamela*, Feld., Reise d. Nov., Lep., p. 199, t. 36, f. 5.

   Hab. Valley of Chillo (8500-9000 feet); Ibarra (7300 feet). Seven examples.

   The black border of the wings of the males is as broad as in specimens from Mexico and Guatemala. In typical *T. gaugamela* these borders are much narrower, but there seems to be every gradation in this respect between the extreme forms.

74. *Terias æquatorialis*, Feld., Wien Ent. Monatschr., v, p. 85.

   Hab. Valley of Chillo (8600 feet). One example.

75. *Terias constancia*, Feld., Reise d. Nov., Lep., p. 200.

   Hab. Nanegal (alt. uncertain). Three examples.

76. *Terias* sp.?

   Hab. Hacienda of Guachala (9217 feet). One example.

   This specimen is too much rubbed to admit of determination. It belongs to the same group as the last.

77. *Pieris xanthodice*, Lucas, Rev. Zool., 1852, p. 337.

   Hab. Valley of Chillo (9000 feet); Machachi (9500-10,000 feet); between Antisanilla and Piñautura (11,000 feet); Pichincha (11-12,000 feet); Illiniza (13,000 feet); Hacienda of Antisana (13,300 feet); Cayambe mountain (13-13,500 feet); Fourth camp, Chimborazo (14,350 feet); south side of Chimborazo (15,000 feet). Very numerous examples.

   These specimens agree with others from different parts of the Andes, the females being somewhat variable, some darker than others.

78. *Pieris marana*, Doubl., Ann. and Mag. N. H., xiv, p. 421; Hew., Ex. Butt., Pieris, t. 6, f. 42.

   Hab. Bridge of Chimbo (1000 feet). One example.

79. *Pieris euthemia*, Feld., Reise d. Nov., Lep., p. 177.

   Hab. Nanegal (alt. uncertain). One example.

80. *Pieris dodia*, Boisd., Sp. Gén., p. 529; Lucas, in La Sagra Hist. Fis. y Pol. Ins. Cuba, p. 492, t. 15, f. 3, 3 a.

   Hab. In the town of Ibarra (7300 feet). Numerous examples.

81. *Pieris suadella*, Feld., Reise d. Nov., Lep., p. 179.

   Hab. Machachi (9800 feet). Two examples.

82. *Pieris cleone*, Doubl. and Hew., Gen. Diurn., Lep., t. 6, f. 6.
Hab. Valley of Chillo (8600 feet). Two examples.

83. *Pieris margarita* (Hübn.).
Mylothris margarita, Hübn., Samml. ex. Schm., ii, t. 120.
Hab. Nanegal and the country west of Quito. Two examples.

84. *Pieris* sp. ?
Hab. Pichincha (12,000 feet). One example.
A much rubbed female specimen of a Pierid with which we are wholly unacquainted. It is in much too damaged a condition to describe.

85. *Callidryas philea* (Linn.).
Papilio philea, Linn., Syst. Nat., i, p. 764.
Hab. Country west of Quito. One example.

86. *Callidryas eubule* (Linn.).
Papilio eubule, Linn., Syst. Nat., i, p. 764.
Hab. Country west of Quito. One example.

87. *Callidryas argante* (Fabr.).
Papilio argante, Fabr., Syst. Ent., p. 470.
Hab. Nanegal (alt. uncertain). Two examples.

88. *Callidryas rurina*, Feld., Reise d. Nov., Lep., p. 194, t. 26, f. 9-11.
Hab. Nanegal (alt. uncertain). Two examples.

89. *Meganostoma cesonia* (Stoll.).
Papilio cesonia, Stoll., Suppl. Cram., t. 41, f. 2, 2b.
Hab. Otovalo (8600 feet). One example.

90. *Colias alticola*, sp. n.
C. *lesbiæ* quoad alarum colores forsan proxima sed multo minor et posticis plaga ad basin costæ vi ulla, alis sulphureis haud rubro-aurantiacis; subtus anticis area discali sulphureis maculis submarginalibus nullis. Exp. 1·7 inches.
♀ mari similis sed pallidior, anticis ad apicem trimaculatis. Femina altera fere albida.
Hab. Pichincha (12,000 feet); Tortorillas, Chimborazo (13,000 feet); Cayambe (13,000 feet); Chimborazo, west side (15,000 feet); Antisana, west side (16,000 feet). Nine examples.
It is with considerable reluctance that we add another species to this complicated genus, but we are unable to associate Mr. Whymper's specimens

with any known species, and Mr. Elwes in his additional notes on the genus *Colias* (Trans. Ent. Soc., 1884, p. 12) in his note on these same specimens, was equally unsuccessful. We have little doubt that of described species it is with *Colias lesbia* that the present insect should be compared, but from it *C. alticola* may be recognised without much difficulty by the points to which we have already drawn attention. A still more nearly allied *Colias* is found in Bolivia, where the late Mr. Buckley took several examples, and it may at some future time be necessary to describe this latter insect. From *C. dimera C. alticola* may be readily distinguished by both the wings being of the same tint, while in the former the primaries are orange and the secondaries pale yellow.

91. *Colias lesbia* (Fabr.).

Papilio lesbia, Fabr., Syst. Ent., p. 477.

Colias lesbia, Elwes, Trans. Ent. Soc., 1884, p. 13.

Hab. Between Otovalo and Ibarra (8000 feet); Otovalo (8600 feet); Valley of Chillo (8600-9000 feet); Machachi (10,000 feet). Fifteen examples.

These specimens agree very fairly, as Mr. Elwes has already pointed out, with others from the Argentine Republic; excepting only that the rosy tint on the upper surface of the wings is almost entirely wanting. The females are all of one type, there being none of the pale form amongst them; but we have little doubt that the insect described by Mr. Kirby as *C. dinora* (Trans. Ent. Soc., 1881, p. 358) should be referred to this pale form. Mr. Henley Smith has kindly lent us the type of *C. dinora*, and we have no doubt upon this point.

92. *Colias dimera*, Doubl. and Hew., Gen. Diurn. Lep., t. 9, f. 3 ; Elwes, Trans. Ent. Soc., 1880, p. 137.

Colias dimera, var. *semperi*, Strecker, Lep.-Rhop. and Het., p. 27, t. 4, f. 4.

Hab. Guallabamba (7200 feet); between Otovalo and Ibarra (8000 feet); Otovalo (8500 feet); Valley of Chillo (9000 feet); Hacienda of Guachala (9217 feet); between Cayambe and Otovalo (9500 feet); Machachi (10,000 feet); between Antisanilla and Piñantura (11,000 feet); Pichincha (11-12,000 feet); Cotocachi (12,000 feet); Illiniza (12-13,000 feet). Very numerous examples.

The females in Mr. Whymper's collection are all of the form described and figured by Mr. Strecker as *Colias semperi*.

APPENDIX—LEPIDOPTERA. 109

Sub-fam. PAPILIONINÆ.

93. *Papilio protesilaus*, Linn., Mus. Ulr., p. 209.
    Hab. Nanegal (alt. uncertain). Two examples.

94. *Papilio iphidamas*, Fabr., Ent. Syst., iii, p. 17.
    Hab. Nanegal (alt. uncertain). One example.

95. *Papilio americus*, Kollar, Denk. Ak. Wien, Math., Cl. i, p. 354.
    Papilio sadalus, Lucas, Rev. Zool., 1852, p. 133, t. 10, f. 4.
    Hab. Machachi (9800-10,000 feet). Five examples.

96. *Papilio thrason*, Feld., Reise d. Nov., Lep., p. 74.
    Hab. Nanegal (alt. uncertain). Two examples.

97. *Papilio pandion*, Feld. ? Reise d. Nov., Lep., p. 79.
    Hab. Country west of Quito. One example.

This specimen has a spot at the end of the cell of the primaries, as in the Central American *P. pandion*, and it agrees generally with examples of that species. But we have never hitherto seen *P. pandion* from so far south, and a single specimen of a species of this difficult group is not sufficient to enable us to speak positively respecting its correct name.

Fam. HESPERIIDÆ.

98. *Thymele eurycles* (Latr.).
    Papilio eurycles, Latr., Enc. Méth., ix, p. 730.
    Hab. Country west of Quito. One example.

99. *Pyrrhopyga* sp. ?
    Pyrrhopyga vulcanus, Hew., Ex. Butt., t. Pyrrhopyga, i, f. 2. (nec Cramer).
    Hab. Country west of Quito. One example agreeing with the figure above referred to and many specimens in our collection. We have not yet found a name for it.

100. *Proteides dalmani* (Latr.).
    Hesperia dalmani, Latr., Enc. Méth., ix, p. 747.
    Hab. Nanegal (alt. uncertain). A single example, in bad condition, agrees with specimens thus named in our collection.

101. *Pamphila phylaeus* (Drury).

Papilio phylaeus, Drury, Ill. Ex. Ent., i, t. 13, f. 4. 5.

Hab. Machachi (9-10,000 feet). Several specimens.

102. *Ancyloxypha* sp. ?

Hab. Valley of Chillo (9-10,000 feet); Machachi (9500-10,000 feet); Quito (9400 feet). Four examples closely allied to *A. melaneura* of Felder, but differing in the absence of the silvery streaks on the secondaries beneath.

103. *Pyrgus orcus* (Cram.).

Papilio orcus, Cram., Pap. Exot., t. 334, f. K. L.

Hab. Bridge of Chimbo (1000 feet). Two examples, which appear referable to this species.

104. *Pythonides tryxus* (Cram.).

Papilio tryxus, Cram., Pap. Exot., t. 334, f. G. H.

Hab. Country west of Quito. One example.

105. *Achlyodes* sp. ?

Hab. Nanegal (alt. uncertain). A single example in bad condition. Agrees with many specimens in our collection from various parts of Central America, but we have not yet found a name for it.

RHYNCHOTA.[1]

By W. L. DISTANT.

Suborder HEMIPTERA-HETEROPTERA.

Fam. PENTATOMIDÆ.

Subfam. CYDNINÆ.

1. *Geotomus nigrocinctus*, Sign. Ann. Soc. Ent. Fr. (1883), p. 40, 8, t. xvii, f. 148.

Hab. Pacific slopes (below 1400 feet). A single example. This species has only recently been described from a specimen in the Vienna Museum with the habitat "Brazil."

Subfam. PENTATOMINÆ.

2. *Thyanta perditor.* *Cimex perditor*, Fabr., Ent. Syst., iv, p. 102, 90 (1794). *Thyanta perditor*, Stål, En. Hem., ii, p. 34, 1 (1872); Dist., Biol. Centr. Am. Rhynch., i, p. 66, 1 (1880).

Hab. Machachi (9-10,000 feet). A single example.

This is not only a variable, but an extremely widely distributed species. Its variable nature has caused it to be described under many different names, all of which will be found in the synonymy of Stål and myself as referred to above. The one specimen collected by Mr. Whymper has the ground colour bright pale olivaceous green, and has the lateral angles of the pronotum spinously produced (this is an inconstant character in the species). It is both a Nearctic and Neotropical species. In North America it has been

---

[1] It was found necessary to publish Mr. Distant's contribution to the *Supplementary Appendix to Travels amongst the Great Andes of the Equator* upon December 17, 1886, in advance of the volume. It is stated, however, that this paper contains errors; and, as it has not been found possible either to obtain corrections of these errors, or the return of the specimens upon which the descriptions were founded, the descriptions are now omitted.—*E. W.*

recorded from Nebraska, Colorado and Texas, it is common in Mexico and Central America, and has been collected both in Colombia and Brazil; whilst in the Antilles, the Islands of Cuba, St. Domingo, and St. Vincent are undoubted habitats.

3. *Arocera splendens*. *Pentatoma splendens*, Blanch., Hist. Nat. Ins., iii, 148, 5 (1841). *Arocera splendens*, Stål, En. Hem., ii, p. 38, 6 (1872); Dist., Biol. Centr. Am. Rhynch., i, p. 75, 7, t. vii, f. 13, 14 (1880).

Hab. Guayaquil (indoors). Two examples.

This is a moderately common Neotropical species. It has not been recorded farther north than Mexico, and extends throughout Central America to Colombia and Venezuela.

4. *Nezara nebulosa*, n. sp.

Hab. Forests above the Bridge of Chimbo (1-3000 feet).

Closely allied to *N. stictica*, Dall., but differing by its very much smaller size, somewhat darker coloration above, the absence of the central longitudinal series of spots to the abdomen beneath, etc.

Long. 12 to 13 millim.

5. *Piezodorus Guildingii*. *Rhaphigaster Guildinii*, Hope, Cat., i, p. 31 (1837). *Piezodorus Guildinii*, Stål, En. Hem., ii, p. 45, 2 (1872); *P. Guildingi*, Dist., Biol. Centr. Am. Rhynch., i, p. 81, 1, t. vii, f. 6 (1880).

Hab. Chillo (9000 feet). A single example. This species is moderately abundant in Central America, has been recorded from the islands of Cuba and St. Vincent, and is probably found throughout the tropical parts of the Neotropical region.

Fam. COREIDÆ.

Division *Spartoceraria*.

6. *Sephina culta*, n. sp.

Hab. Milligalli (6200 feet). A single example.

Long. 22 millim. Exp. lat. ang. pron. 8 millim.

This species, by its peculiar markings, is nearest allied to *S. geniculata*, Dist., received from Costa Rica.

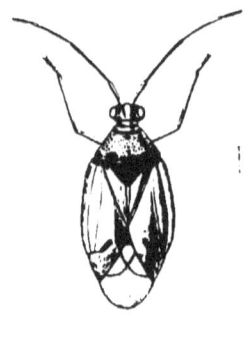

LYDE TRANSLUCIDA, DISTANT.
PICHINCHA, 12,000 FEET.

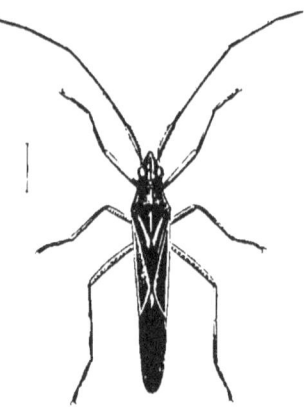

NEOMERIS PRÆCELSUS, DISTANT.
HACIENDA OF ANTISANA, 13,300 FEET.

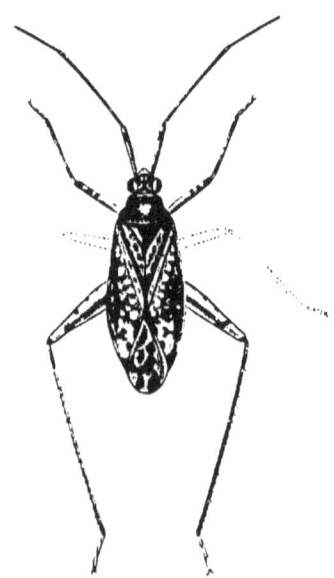

CONYZA VARIEGATA, DISTANT.
EASTERN SIDE OF CORAZON, 12,000 FEET

Division *Corcaria*.

7. *Margus tibialis*, n. sp.

Hab. Eastern side of Corazon (12,000 feet), eastern side of Pichincha (12,000 feet), Hacienda of Guachala (9217 feet), Machachi (9-10,000 feet), Pacific slopes (7-8000 feet). Six examples.

Long. 7 to 8 millim.

This species is allied to both *M. pectoralis*, and *M. pallipes* of Dallas.

Division *Harmostaria*.

8. *Harmostes Corazonus*, n. sp.

Hab. Eastern side of Corazon (12,000 feet). Two examples.

Long. $5\frac{1}{2}$ millim.

This species is apparently allied to the Chilian *H. raphimerus*, Spin.

9. *H. montivagus*, n. sp.

Hab. Machachi (9-10,000 feet), eastern side of Corazon (12,000 feet). Three examples. Long. 6 millim.

Fam. *LYG.EIDÆ*.

Division *Orsillaria*.

10. *Nysius procerus*, n. sp.

Hab. Machachi (9-10,000 feet). One example.

Long. 4 millim.

This species should be nearest allied to the Colombian *N. nubilus*, Dall., the type of which is no longer to be found in the British Museum.

Fam. *CAPSIDÆ*.

Subfam. CAPSINÆ.

Division *Miraria*.

NEOMIRIS, gen. nov.

11. *Neomiris præcelsus*, n. sp.

Hab. Hacienda of Antisana (13,300 feet). Two examples.

Long. 8 millim. For Figure see the accompanying Plate.

Division *Phytocoraria.*

DIONYZA, gen. nov.

12. *Dionyza variegata,* n. sp.
Hab. Eastern side of Corazon (12,000 feet). A single example.
Long. 7 millim. For Figure see the Plate facing page 113.

13. *Calocoris montanus,* n. sp.
Hab. La Dormida, Cayambe (11,800 feet), Pacific slopes (7-8000 feet). Seven examples. Long. 7 millim.

Only one perfectly developed specimen of this species was captured by Mr. Whymper, the others being immature specimens.

Division *Capsaria.*

14. *Lygus collinus,* n. sp.
Hab. Hacienda of Guachala (9217 feet). Two examples.
Long. 6 millim.

15. *L. sublimatus,* n. sp.
Hab. La Dormida, Cayambe (11,800 feet). A single example.
Long. 5 millim.

16. *L. excelsus,* n. sp.
Hab. Eastern side of Corazon (12,000 feet). Two examples.
Long. 5 millim.

Division *Bryocoraria.*

LYDE, gen. nov.

17. *Lyde translucida,* n. sp.
Hab. Pichincha (12,000 feet). A single example. Long. 4½ millim.
For Figure see the Plate facing page 113.

Fam. *ARADIDÆ.*

Subfam. BRACHYRHYNCHINÆ.

Division *Brachyrhyncharia.*

18. *Cinyphus? obscurus,* n. sp.
Hab. Forests above the Bridge of Chimbo (1-3000 feet). A single example.
Long. 8 millim.

I have provisionally retained this species in the genus *Cinyphus*, to which it has the strongest affinities. It differs, however, in the structure of the

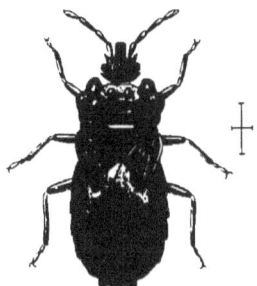

CINYPHUS? OBSCURUS, DISTANT.
FORESTS ABOVE THE BRIDGE OF CHIMBO.

antennæ, and will doubtless eventually necessitate the creation of a new genus for its reception.

19. *Aneurus flavomaculatus*, n. sp.

Hab. Eastern slopes of Pichincha (12,000 feet). Nine examples.

Long. 6 millim.

Fam. REDUVIIDÆ.

Subfam. REDUVIINÆ.

20. *Prionotus carinatus*. *Cimex carinatus*, Forst., Nov. Spec. Ins., p. 72, 72 (1771). *Prionotus carinatus*, Stål, En. Hem., ii, p. 72, 2 (1872).

Hab. Nanegal (3-4000 feet). A single example.

This species may be considered as not extending much farther north than Ecuador. In Colombia it is replaced by *P. gallus*, Stål. *P. carinatus* is a Brazilian species, and is common in the neighbourhood of Rio Janeiro. I have also received it from Paraguay.

Subfam. ACANTHASPIDINÆ.

21. *Conorhinus dimidiatus*. *Reduvius dimidiatus*, Latr., in Humb. & Bonpl. Obs. Zool., i, p. 149, t. 15, f. 11. *Conorhinus dimidiatus*, Stål, Berl. Ent. Zeitschr., iii, p. 110, 7 (1859).

Hab. Guayaquil (indoors). Five examples.

This species extends as far north as Mexico, and is common in Central America. It had previously been recorded from Guayaquil by Stål.

116   TRAVELS AMONGST THE GREAT ANDES.

22. *Conorhinus* sp.?

Hab. La Mona (100 feet). A single example.

I am in doubt as to the identity of this species, and in the absence of typical examples of those described by Philippi, refrain from describing it as a new species.

Subfam. STENOPODINÆ.

23. *Stenopoda scutellata,* n. sp.

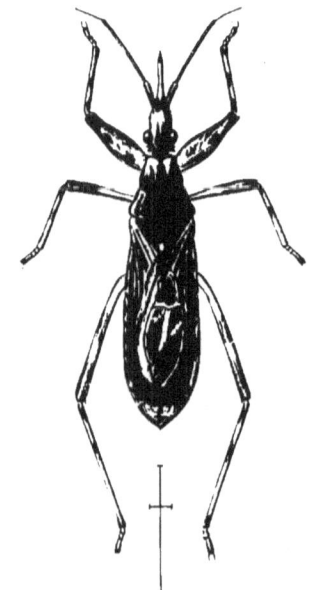

STENOPODA SCUTELLATA, DISTANT.
GUAYAQUIL.

Hab. Guayaquil (indoors). Two examples. Long. 20 millim.

APPENDIX—RHYNCHOTA.

24. *Pnohirmus Whymperi*, n. sp.

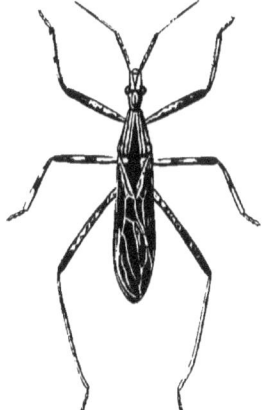

PNOHIRMUS WHYMPERI, DISTANT.
LA MONA.

Hab. La Mona (100 feet). Three examples.
Long. 13 to 14 millim.

Subfam. EMESINÆ.

25. *Emesa* sp.

Hab. Illiniza (16,500 feet).

A single specimen of this genus was captured by Mr. Whymper at this elevation, and was, as that traveller informs me, "the highest animal life of any kind obtained or observed. I did not even see Condors so high as this." The specimen being contained in spirit, was incapable of exact identification, though I dissected it for that purpose. Its nearest ally is the *E. longipes*, De Geer, a well-known North American insect which is described as living "in the branches of small pine trees, and in outhouses and barns."

Fam. ACANTHIADÆ.

26. *Acanthia Andensis*, n. sp.

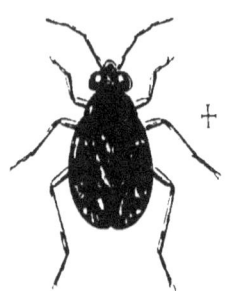

ACANTHIA ANDENSIS, DISTANT.
MACHACHI, 10,000 FEET.

Hab. Machachi (9-10,000 feet). A single example.
Long. 4 millim.

Fam. BELOSTOMIDÆ.

27. *Zaitha anura.* *Diplonychus anurus*, H.-S., Wanz. Ins., viii, p. 26, t. 257, f. 799 (1848). *Zaitha anurus*, Mayr, Verh. d. Zool.-bot. Ges. Wien, xxi, pp. 408 and 412, 6 (1871).

Hab. Guayaquil (indoors). Two examples.

A very widely distributed species, ranging from Mexico to the Argentine Republic, and recorded from Cuba.

Suborder HEMIPTERA-HOMOPTERA.

Fam. CICADIDÆ.

28. *Zammara smaragdina*, Walker, List. Hom., i, p. 33, 3 (1850); *ib.*, iv, t. 1, f. 4 (1852); Dist., Biol. Centr. Am. Rhynch. Hom., p. 3, 1, t. 1, f. 1, 1a, 1b (1881).

Hab. Nanegal (3-4000 feet).

This species is found in Mexico, is moderately abundant throughout Central America, and I have already examined specimens from Colombia, Venezuela, and Ecuador, contained in the Dresden Museum.

29. *Carineta socia*, Uhler, Proc. Bost. Soc. Nat. Hist., xvii, p. 285 (1875).
Hab. La Mona (100 feet), Tanti (1890 feet). Two examples.

This species was described from specimens collected on the Lower Amazons, and I have since also examined specimens collected by Moritz in Colombia.

30. *C. basalis*, Walker, List. Hom., i, p. 245, 7 (1850).
Hab. Nanegal (3-4000), Chillo (9000 feet). Numerous examples.

Hitherto recorded from Colombia and Venezuela.

31. *C. fimbriata*, Walker, MS.
Hab. Nanegal (3-4000 feet), Quito (9350 feet), Machachi (9-10,000 feet). Three examples.

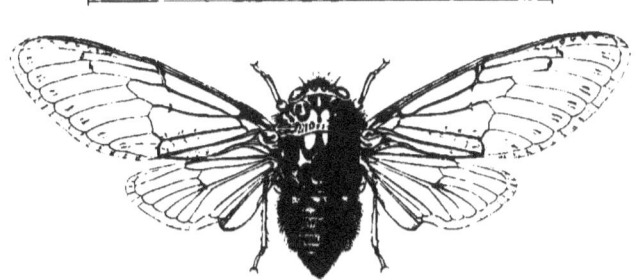

CARINETA FIMBRIATA, WALKER.
MACHACHI, 10,000 FEET.

This species is named *C. fimbriata*, Walk., in the collection of the British Museum, but I have failed to find any published description of the species. Long. 21 millim. Exp. tegm. 70 millim.

I had previously received this species from Ecuador, where it seems to be of a somewhat abundant character.

Fam. CERCOPIDÆ.
Subfam. CERCOPINÆ.

32. *Sphenorhina ruida*, n. sp.
Hab. Forests above the Bridge of Chimbo (1-3000 feet). A single example. Long. 8 millim.

33. *S. Jullia*, n. sp.
Hab. Forests above the Bridge of Chimbo (1-3000 feet). A single example. Long. 9 millim.

Fam. MEMBRACIDÆ.

Subfam. SMILIINÆ.

34. *Heranice miltoglypta.* Thelia miltoglypta, Fairm., Ann. Soc. Ent. Fr., ser. 2, vol. iv, p. 306, 2, t. 5, f. 4, 12 (1846).

Hab. Machachi (9-10,000 feet), Corazon (12,000 feet). Three examples. This is an abundant species in Colombia.

35. *Acutalis* sp.

Hab. Pichincha (12,000 feet). A somewhat mutilated specimen, which may be an undescribed species, and is allied to *A. terminalis*, Walk.

Fam. JASSIDÆ.

Subfam. TETTIGONIINÆ.

36. *Tettigonia Medusa*, n. sp.

Hab. Machachi (9-10,000 feet). A single example. Long. 8 millim. This species is allied to *T. Walkeri*, Sign., a species received from Quito.

37. *T. duplicaria*, n. sp.

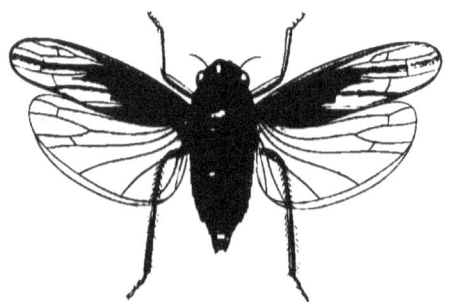

TETTIGONIA DUPLICARIA, DISTANT.
MACHACHI, 10 000 FEET.

Hab. Machachi (9-10,000 feet), Hacienda of Guachala (9217 feet). Four examples. Long. 8 millim.

38. *Tettigonia* sp.

Hab. Forests above the Bridge of Chimbo (1-3000 feet). One specimen collected by Mr. Whymper is closely allied to, if not identical with, *T. pruinosa*, Walk.

CRUSTACEA.

PODOPHTHALMIA.

By EDW. J. MIERS, F.L.S., F.Z.S.

The *Podophthalmia* collected by Mr. Whymper in Ecuador comprise only some half-dozen species, but among these the specimens of *Pseudothelphusa macropa* are of special interest, on account of the high altitude at which they were obtained. The small Crayfish which, according to Prof. Orton, abounds in the stagnant waters about Quito,[1] does not occur in Mr. Whymper's collection.

Upon the whole series obtained it may be observed, that the species which have rewarded Mr. Whymper's search are less numerous than might with reason have been anticipated. Even if account be taken solely of terrestrial and fluviatile forms, some additional *Thelphusidea* or terrestrial *Palaemonidae* might have been expected to occur in the Collection. It should, however, in justice to Mr. Whymper, be noted that Prof. Orton (l.c.) says the Crayfish above referred to is, he believes, the only Crustacean inhabiting the Quito valley. Species of this class are therefore, probably, scarce.

1. *Pseudothelphusa macropa*. (Figs. **A**, **B**.)

Boscia macropa, M. Edwards, Ann. Sci. Nat. (3me série), Zool. xx, p. 208 (1853); Arch. Mus. d'Hist. Nat., vii, p. 175, pl. xii, fig. 3 (1854).

Pseudothelphusa macropa, S. I. Smith, Trans. Connect. Acad., ii, p. 146 (1870).

Hab. Milligalli (6200 feet); Plain of Tumbaco (7850 feet). Five examples.

Four small females are referred to this species, obtained at Milligalli, which according to Mr. Whymper's information is about 39 miles by road from Quito, and about 6000 feet above the sea ; and also an adult but small female reported to have been taken on the Plain of Tumbaco, about three hours north of Quito, and 7850 feet above the sea. This, the largest example, measures as follows :—Ad. ♀, length of carapace about 7 lines (15 millim.); breadth of carapace about 12 lines (25 millim.). I subjoin a description of these specimens.

Carapace transverse, its greatest width about 1½ times exceeding its length ; depressed and nearly flat on the dorsal surface, which is finely

[1] American Naturalist, vi, p. 650 (1872).

punctulated and but slightly convex near to the frontal and anterior margin; the mesogastric and cervical sutures are faintly defined; the antero-lateral margins are regularly and strongly arcuated, and are granulated rather than denticulated. The front is deflexed, less than ⅓ the width of the carapace, and its anterior margin is sinuated, and posterior to it there is sometimes a transverse elevation, which is indistinctly granulated (see Fig. A). The orbital margins are without fissures, and are rather indistinctly granulated. The pterygostomian regions are nearly smooth. The eye peduncles are short, and not

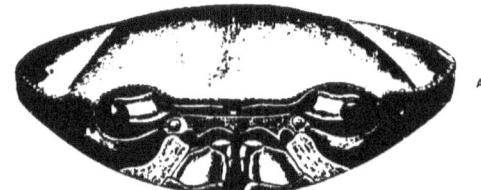

PSEUDOTHELPHUSA MACROPA, M. EDWARDS.

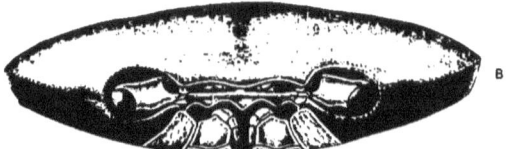

PSEUDOTHELPHUSA MACROPA, VAR. PLANA, S. I. SMITH (?).

very stout, and do not quite fill the orbital cavities. The exterior maxillipedes are punctulated and are of the form figured by Milne Edwards as characteristic of *P. macropa*. The chelipedes are small and unequal, the merus in the larger (left) chelipede is short, trigonous, scarcely reaching beyond the antero-lateral margins of the carapace; its anterior margin is armed with small, somewhat triangular teeth, its upper surface has the smooth, oval depression (very slightly indicated) which is referred to by Milne Edwards in his description of *P. macropa*; the carpus has an acute spine on the interior margin; the palm is slightly convex, rounded above, about as long as the merus and finely granulated, fingers about as long as the palm, with the apices slightly incurved, acute, and the interior margins armed with regular and rather small teeth. In the smaller chelipede the palm is rather shorter, and the fingers are straighter. The ambulatory legs (which are imperfect) are of moderate length, and but slightly compressed, and naked, the dactyli styliform slightly longer than the penultimate joints, nearly straight, and armed with rather distant spinules.

In the specimens from Milligalli the punctulations of the carapace are much more distinct and numerous, the granulated post-frontal ridge is obsolete, and the eyes more nearly fill the orbital cavities (see Fig. B). In the single example from Tumbaco, in which the rudimentary granulated postfrontal crest is slightly developed, the carapace is minutely granulated and the eye peduncles are rather slenderer.

It will be observed that the specimens differ in some particulars from Milne Edwards' description, particularly as regards the antero-lateral margins of the carapace, which although not denticulated are very distinctly granulated.

The specimens from Milligalli may very probably be identical with the form designated by S. I. Smith, *Pseudothelphusa plana*, and may be specifically distinct; the largest female, which is nearly the same size as the specimen from Tumbaco, is proportionately slightly narrower, measuring as follows :—
♀. Length of carapace about 7 lines (15 millims.); breadth of carapace nearly 12 lines (24·5 millims.). There are examples of both varieties, from Guatemala, in the Collection of the British (Natural History) Museum.

2. *Aratus Pisoni*.

Sesarma Pisoni, M. Edwards, Hist. Nat. Crust., ii, p. 76, pl. xvi, figs. 4, 5 (1837).

Aratus Pisoni, M. Edwards, Ann. Sci. Nat. (3me série), Zool. xx, p. 187 (1853); Kingsley, Proc. Acad. Nat. Sci. Philad., p. 218 (1880).

Hab. Guayaquil. A single female specimen. Length of carapace about 6½ lines (13·5 millims.); breadth of carapace about 6½ lines (13·5 millims.).

The specimen is a small one and somewhat faded; but it cannot, I think, be distinguished specifically from specimens from Jamaica and Brazil in the Collection of the British Museum. The occurrence of this species on the Western Coast (at Nicaragua) has already been noted by Kingsley. On the Eastern Coast its range extends southwards to Rio de Janeiro, Brazil (Heller).

At Guayaquil two Crabs were collected with *Aratus Pisoni* which probably belong to different species of the genus *Gelasimus*, but their identification must remain uncertain since one is a female, and the other, although a male, has lost the larger chelipede. In both of these specimens the front between the eyes is broad, and the antero-lateral margins are anteriorly nearly parallel, and convergent from a point placed at some distance behind the exterior angles of the orbits, as in several American species of this genus.

3. *Palæmon Jamaicensis.*
    Cancer (Astacus) Jamaicensis, Herbst, Naturgesch. der Krabben u. Krebse, ii (heft 2), p. 57, pl. xxvii, fig. 2 (1792).
    Palæmon Jamaicensis, Olivier, Encycl. Méth., viii, p. 659 (1811); M. Edwards, Hist. Nat. des Crust., ii, p. 398 (1837); v. Martens, Arch. f. Nat., xxxv, p. 23 (1869); xxxviii, p. 137 (1872); S. I. Smith, Rep. Peab. Acad. Sci., p. 97 (1869), publ. 1871. Trans. Connecticut Acad., ii, p. 23 (1869); Kingsley, Bull. Essex Institute, x, p. 68 (1878); xiv, p. 107 (1883).
    Palæmon brachydactylus, Wiegmann, Arch. f. Naturgesch., ii, p. 148 (1836), var. ?
    Palæmon punctatus, Randall, Journ. Ac. Nat. Sci. Phil., p. 146 (1839).
    Macrobrachium Americanum, Spence Bate, Proc. Zool. Soc., p. 363, pl. xxx (1868); c.f. Semper, Proc. Zool. Soc., t. c. p. 585 (1868).
    Hab. Tanti (1890 feet). One example.
    The specimen, which is adult, has the rostrum broken, but its identification is I think certain.
    ♂. Length of body to base of rostrum $3\frac{1}{3}$ inches (85 millim.).
    This species occurs commonly in the West Indies and Brazil, and in the fresh waters of Mexico and Central America, and Prof. Kingsley has already recorded its having been brought by Prof. Orton from the junction of the Napo and Maranon Rivers, and from Guatemala, and from Polvon, W. Nicaragua. There are small specimens which I think are referable to *P. Jamaicensis,* from the Cape Verde Islands, in the Collection of the British (Natural History) Museum.

4. *Squilla dubia.* (See the accompanying Plate.)
    Squilla dubia, M. Edwards, Hist. Nat. Crust., ii, p. 522 (1837); Miers, Ann. Mag. Nat. Hist. (ser. 5), v, p. 24 (1880) and *synonyma.*
    Hab. Guayaquil. Two examples.
    Two adult males are in the collection, obtained from the saline backwater at Guayaquil. In all essential particulars, they resemble specimens from San Domingo and Honduras in the Collection of the British (Natural History) Museum. Mr. Whymper was informed that the local name was 'Camaron brujo.' "It lives in the mud, and the natives have tried to eat it, but found it poisonous."
    Ad. ♂. Length of body to end of rostrum $5\frac{3}{4}$ inches (147 millim.).
    As Milne Edwards' original description is very short, and is unaccompanied by any illustration, it has been thought well to figure the smaller and more perfect male collected by Mr. Whymper, in order to facilitate the identification of this species, which if not correctly referred to *Squilla dubia* must be designated by Dana's name, *S. rubrolineata.*

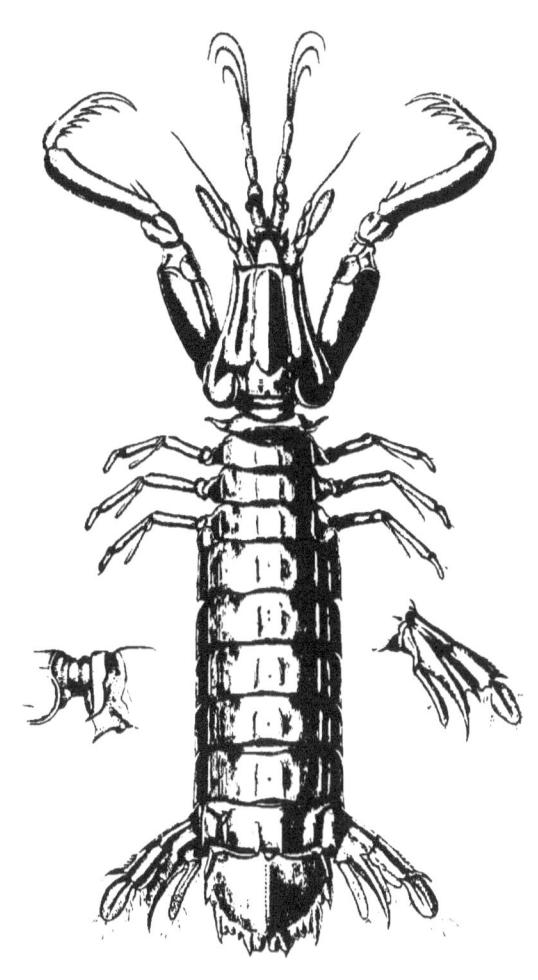

SQUILLA DUBIA (?), M. EDWARDS.
FROM THE SALINE BACKWATER AT GUAYAQUIL.

CRUSTACEA—(Continued).

BY THE REV. A. E. EATON, M.A.

ISOPODA.

The species of Terrestrial *Isopoda* collected by Mr. Whymper in Ecuador are three in number, represented altogether by twenty-five specimens, captured in eight localities, mostly at high altitudes. One of these species was previously known as a native of Chili; the others are met with in nearly all countries.

1. *Philoscia angustata*, Nicolet [in *Oniscus*, Nicolet], Gay's Hist. Fis. y Polit. de Chile, Zool., iii, 268, Atlas Crustaceos, iii, 8-8b (1849).

Hab. Lower slopes of Pichincha (12,000 feet); La Dormida, Cayambe (11,800 feet). Five examples.

2. *Porcellio lævis*, Latreille, Hist. Nat. d. Crust. & Ins., vii, 46 (1804); Lereboullet, Mém. Soc. Mus. d'Hist. Nat. Strasbourg, iv, 45-49, pl. i, 7 and (details) iii, 55-60 (1853).

Hab. Among stones at Quito (9400 feet). Twelve examples.

3. *Metoponorthus pruinosus*, Brandt [in *Porcellio*, Bdt.], Bull. Soc. Imp. Nat. Moscou, vi, 181 (1833).

Hab. Hacienda of Antisana (13,300 feet); track between Antisanilla and Piñantura (11,000 feet); lower slopes of Pichincha (12,000 feet); Hacienda of Guachala (9200 feet); garden of Señor Gomez de la Torre, Ibarra (7200 feet); and Guayaquil. Eight examples.

BY THE REV. T. R. R. STEBBING, M.A.

AMPHIPODA.

1. *Hyalella inermis*, S. I. Smith, in Ann. Rep. U.S. Geol. and Geograph. Survey of the Terr., 1873, part iii, Zool., p. 609, Amphipods pl. I, figs. 1-2. [Washington, 1875.]

Hab. Hacienda of Antisana (13,300 feet); Valley of Collanes, Altar (12,500 feet); Machachi (9800 feet). Numerous examples, male and female.

Mr. Whymper has given me the following details of the capture of this *Amphipod* at the above-named localities; which, in view of the great heights named, are worthy of record. He says :—" The specimens from Machachi were the first crustacea I obtained in Ecuador, and were taken with my own hands from a tiny rivulet, in a stagnant place, to the S.E. of the town. I have a long series of observations of mercurial barometer for the height of Machachi, and regard the deduced altitude as one of the best obtained on the journey."

"The specimens from the Valley of Collanes, and those from the vicinity of the Hacienda (farm) of Antisana (situated on the lower slopes of Antisana) came in each case from small pools. They were taken by my European assistants and brought to me on the spot. The heights given for these localities do not depend upon a single observation of mercurial barometer, but upon several observations in each case, and the determinations are perhaps within 50 feet of the truth."

In the Report this species is said to have been taken in Colorado. Mr. Walter Faxon says that it was "collected by Mr. Agassiz at San Antonio, Peru, in saline water, 3300 feet above the sea; nitrate district of Pisagua," and "during the voyage of the 'Hassler' at Puerto Bueno, Smyth Channel, Straits of Magellan." He also thinks that it should be united with the very widely distributed species *Hyalella dentata*, S. I. Smith, under the name *Allorchestes dentatus*, var. *inermis*. There are, however, to my mind sufficient reasons for retaining both the generic and specific names given it by Mr. S. I. Smith, with a reserve in favour of the name *andina* mentioned below.

So far as I am aware, no species of Amphipod has been recorded from heights so great as those of Mr. Whymper's stations here mentioned. *Gammarus lacustris*, now called *Gammarus limnæus*, S. I. Smith, was taken by Lieut. Carpenter in Colorado at an elevation of 9000 feet. In Europe, Heller mentions the capture by Dr. Kotschy of a species called *Orchestia carimana* on Mount Olympus in Cyprus, at a height of 4000 feet, which was thought surprising. Philippi, on his journey through the Chilian desert of Atacama, the account of which was published in 1860, took a species which he calls *Amphithoë andina*, at heights varying from 7500 to 10,500 feet above the sea-level. It may be inferred from his description that he had before him a species of *Hyalella*, and not improbably that which has since been distinguished by the specific name *inermis*. Wrześniowski in 1879 described three species belonging to the same genus, taken at heights of 7000 and 8000 feet above the sea-level, on the western and eastern slopes of the Cordilleras. This author arranges the species in question in the genus *Hyale*, subgenus

Allorchestes, and discriminates the species themselves by characters relating to the branchiæ not noticed in earlier descriptions.

Hyalella inermis received its name to distinguish it from the allied species *Hyalella dentata*, which has the first and second segments of the abdomen or pleon "with the dorsal margin produced posteriorly into a well-marked spiniform tooth." In these species the second gnathopod or clasper is very different in the two sexes, the hand in the male being greatly developed and set in the short transverse wrist; while in the female hand and wrist are in a continuous line, about equal in length, quite small and thin.

REPTILIA & BATRACHIA.

BY G. A. BOULENGER.

An account of the herpetological collection made by Mr. Whymper in Ecuador was published in the Annals and Magazine of Natural History (5) ix, 1882, pp. 457-467. In the following list I have introduced a few corrections to my previous determinations, as well as some changes in the nomenclature of the Lizards, which is now in accordance with the recently published Catalogue of Lizards in the British Museum.

REPTILIA.

CHELONIA.

1. *Cinosternum*, sp.

   Hab. Nanegal (3000 feet). Two very young, dried specimens,[1] the dorsal shield 24 millim. long; too small and too badly preserved to be properly identified.

   These tortoises are closely allied to *C. leucostomum*, A. Dum., which occurs in Colombia; but the axillary and inguinal shields are in contact, as in *C. integrum*, Leconte, from Mexico.

   This is, I believe, the first time that a *Cinosternum* is recorded from Ecuador.

LACERTILIA.

2. *Gonatodes caudiscutatus*, (Gthr.). *Gymnodactylus caudiscutatus*, Günth., Proc. Zool. Soc., 1859, p. 410.

   Hab. Guayaquil. One half-grown specimen.

3. *Anolis Fraseri*, Gthr. *Anolis Fraseri*, part., Günth., Proc. Zool. Soc., 1859, p. 407; *Anolis Perillei*, Bouleng., Bull. Soc. Zool. France, 1880, p. 42.

   Hab. Nanegal (3000 feet). One ♂ specimen.

4. *A. Andianus*, Bouleng. Cat. Liz. ii, p. 60. *Anolis squamulatus?* (non Ptrs.) Bouleng., Ann. and Mag. N. H. (5), ix, p. 458.

   Hab. Milligalli (6200 feet). One ♀ specimen.

   Head twice as long as broad, much longer than the tibia; forehead and

[1] Bought in this condition at Quito.—*E. W.*

interorbital region concave, no frontal ridges; upper head-scales very small, rugose; scales of the supraorbital semicircles scarcely enlarged, separated by five series of scales; ten to twelve enlarged, tricarinate, supraocular scales, separated from the supraorbitals by a series of granules; occipital not enlarged; canthus rostralis feebly marked, no enlarged canthal scales; loreal rows six; seven labials to below the centre of the eye; ear-opening moderate, oval. Gular appendage absent (♀). Body slightly compressed. Scales very small, granular, smooth, equal on the back and sides; ventrals very small, larger than dorsals, granular, smooth. The adpressed hind limb reaches between the ear and the orbit; digits moderately dilated; 18 lamellæ under phalanges II and III of the fourth toe. Tail cylindrical, covered above with small equal feebly-keeled scales; its length almost thrice that of head and body. Dull lilac above, minutely and indistinctly speckled with blackish; lower surfaces whitish.

Total length	230 millim.	Fore limb	25 mm.
Head	18 ,,	Hind limb .	42 ,,
Width of head	9 ,,	Tibia .	12 ,,
Body	47 ,,	Tail	175 ,,

5. *A. stigmosus*, Bocourt, Nouv. Arch. Mus. Paris, v, 1869, Bull. p. 43.

Hab. Tanti (1890 feet). One ♀ specimen.

6. *Liocephalus trachycephalus*, (A. Dum.). *Holotropis trachycephalus*, A. Dum., Cat. Méth. Rept., p. 76.

Hab. Otovalo (8460 feet); between Quito and Guallabamba (8500 feet); between Guallabamba and Guachala (8000 feet); Ambato (8606 feet); Machachi (9-10,000 feet); La Dormida, Cayambe mountain (11,800 feet); and Hacienda de la Rosario, on the lower slopes of Illiniza (10,356 feet). Twenty-five specimens.

7. *L. iridescens*, Gthr. Proc. Zool. Soc., 1859, p. 409, pl. xx, fig. B.

Hab. Guayaquil. One specimen.

8. *Ameiva septemlineata*, A. Dum., Cat. Méth. Rept., p. 112. *Ameiva sexscutata*, Günth., Proc. Zool. Soc., 1859, p. 402.

Hab. Tanti (1890 feet). Two specimens (♀ and h. gr.).

9. *Ecpleopus (Pholidobolus) montium*, Peters, Abh. Berl. Ac., 1862, p. 196, pl. ii, fig. 3.

Hab. Hacienda Olalla (8500 feet); Chillo (9000 feet); and lower slopes of Pichincha (11,000 feet). Nine specimens.

10. *Proctoporus unicolor*, (Gray). *Riama unicolor*, Gray, Proc. Zool. Soc., 1858, p. 446, pl. xv, fig. 2.

Hab. Hacienda Olalla (8500 feet); Chillo (9000 feet). Seven specimens. A small median occipital shield is frequently present.

11. *Amphisbæna fuliginosa*, L. Dum. & Bibr., Erp. Gén., v, p. 480.

Hab. Tanti (1890 feet); Guayaquil. Two specimens.

OPHIDIA.

From the interior of Ecuador Mr. Whymper obtained only two snakes, belonging to two species, viz. *Liophis alticola* and *Leptognathus nebulatus*; and he observes:—"The most intelligent persons I could question declared that snakes did not exist; and the surprise and curiosity which these two specimens excited amongst the natives showed that they were rare." In his paper on the reptiles collected by the Orton expedition Prof. Cope mentions no less than nine species of snakes from the "valley of Quito." This is in contradiction with what Orton himself says:—"During a residence of nearly three months in the Quito valley we saw but one snake" ('The Andes and the Amazon,' English edition, p. 107).

12. *Boa constrictor*, L. Jan, Icon. Ophid., 5ᵉ. livr., pl. ii, fig. 2.

Hab. Guayaquil. One young specimen.

13. *Homalocranion melanocephalum*, (L.). Jan, Icon. Ophid., 13ᵉ. livr., pl. iii, fig. 4.

Hab. Milligalli (6200 feet). One specimen.

14. *Coronella doliata*, (L.), var. *formosa*. Jan, Icon. Ophid., 14ᵉ. livr., pl. iv, fig. 1.

Hab. Guayaquil. One specimen, with undivided anal and twenty-three rows of scales.

15. *C. Whymperi*, Bouleng., Ann. and Mag. N. H. (5) ix, p. 460.

Hab. Milligalli (6200 feet). Two specimens.

Habit of *Coronella austriaca*. Head moderate; snout short, its length not quite twice the diameter of the eye. Rostral moderate, not advancing on the internasals; latter longer than broad; frontal as long as its distance from the tip of the snout, its front edge nearly straight; parietals longer than frontal, narrowed and including a considerable notch behind. Eight supero-labials, fourth and fifth entering the eye; one preocular, two postoculars, lower smaller than upper; a single anterior temporal; eight inferior labials, five in contact with mentals; latter, hinder pair longest. Scales in seventeen rows. Gastrosteges 154 or 156; anal bifid; urosteges 55 or 66. Brown above, upper half of supero-labials yellowish, lower half blackish; a black streak from the eye along the side of the neck; a light black-edged spot on each side of the nape; a rather indistinct, interrupted, yellowish line along each side of the front half of the body, between the fifth and sixth rows of

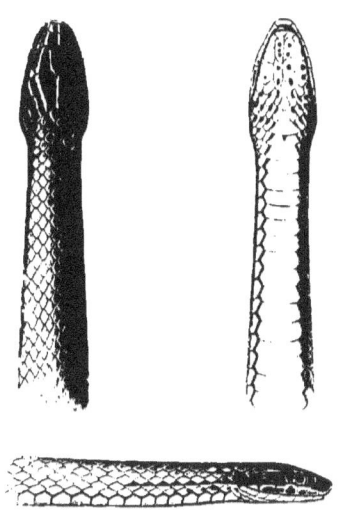

CORONELLA WHYMPERI, BOULENGER.
MILLIGALLI, 6200 FEET.

scales; a black stripe along the middle of the tail and of the hind part of the back; yellow or brownish-yellow beneath; outer edge of gastrosteges and urosteges, and sometimes front edge of former, black.

Length of the two specimens:—Head and body 514, 410 millim.; tail 127, 135 millim.

This species bears a close resemblance to *Coronella decorata*, Gthr. (Cat. Colubr. Sn., p. 35), from Mexico, but differs in the narrower internasals, shorter tail, size, and coloration.

16. *Liophis reginæ*, (L.), var. *albiventris*, Jan, Icon. Ophid., 16ᵉ. livr., pl. vi.

Hab. Milligalli (two adult and two young); Tanti (one half-grown). The var. *quadrilineatus*, Jan, is represented by two specimens, which are evidently the adult state of var. *albiventris*.

17. *L. alticola*, (Cope). *Opheomorphus alticolus*, Cope, Proc. Ac. N. S. Philad., 1868, p. 102.

Hab. Hacienda Olalla, plain of Tumbaco (8500 feet). One fine specimen, which was brought in to Mr. Whymper alive, and excited much curiosity amongst the natives.

18. *L. splendens*, Jan, Icon. Ophid., 18°. livr., pl. v, fig. 1.
Hab. Hacienda of Palmira, Nanegal (about 3000 feet). One specimen.

19. *Herpetodryas brunneus*, Gthr. Cat. Col. Sn., p. 116.
Hab. Guayaquil. One specimen.

20. *Oxyrhopus clælia*, (Daud.). Jan, Icon. Ophid., 35°. livr., pl. i, fig. 1.
Hab. Hacienda of Palmira, Nanegal. One specimen.

21. *O. petolarius*, (L.). Jan, Icon. Ophid., 36°. livr., pl. i, fig. 1.
Hab. Guayaquil. One specimen.

22. *Himantodes cenchoa*, (L.). Dum. & Bibr., Erp. Gén., vii, p. 1065.
Hab. Guayaquil. One specimen.

23. *Leptognathus nebulatus*, (L.). Jan, Icon. Ophid., 37°. livr., pl. v, fig. 3.
Hab. Ibarra (7300 feet). One half-grown specimen, which was brought in to Mr. Whymper alive.

24. *Elaps Maregrari*, Wied, var. *ancolaris*, Jan, Icon. Ophid., 42°. livr., pl. iv, fig. 2.
Hab. Nanegal (3000 feet). One specimen.

25. *E. lemniscatus*, (L.). Jan, Icon. Ophid., 42°. livr., pl. v, fig. 1.
Hab. Guayaquil. One specimen.

26. *Bothrops atrox*, (L.). Jan, Icon. Ophid., 47°. livr., pl. ii.
Hab. Nanegal (3000 feet); Mindo, W. of Quito (about 4000 feet). Three specimens.

27. *B. Schlegeli*, (Berthold). Jan, Icon. Ophid., 47°. livr., pl. vi, fig. 2.
Hab. Bologna (height unknown); S. Domingo de los Colorados (height unknown). Obtained at Quito. Two adult, and two young specimens.

BATRACHIA.

ECAUDATA.

28. *Prostherapis Whymperi*, Bouleng., Ann. and Mag. N. H. (5), ix, p. 462.
Hab. Tanti (1890 feet). A single ♂ specimen.

Snout depressed, projecting, truncate, with angular canthus rostralis;

loreal region nearly vertical; nostril nearer the tip of the snout than the eye; the greatest diameter of the orbit equals the length of the snout; interorbital space broader than the upper eyelid; tympanum perfectly distinct, two-thirds the breadth of the eye. First and second fingers equal; toes quite free; disks of fingers and toes small; subarticular and inner metatarsal tubercles indistinct; no outer metatarsal tubercle. The hind limb being carried forwards along the body, the tibio-tarsal articulation marks the anterior border of the eye. Skin everywhere perfectly smooth. Blackish; throat and belly marbled with grey; lower surface of hind limbs greyish, of arms whitish. An internal subgular vocal sac. From snout to vent 24 millim.

PROSTHERAPIS WHYMPERI, BOULENGER.
TANTI, 1890 FEET.

This small frog is closely allied to *P. inguinalis*, Cope, hitherto the unique species of the interesting genus *Prostherapis*, Cope. The British Museum having recently received the latter from Ecuador (Sarayacu and Canelos) through Mr. Buckley, I have been able to compare it with this new form; and I have no doubt they are perfectly distinct from each other. In *P. inguinalis* the tympanum is hidden, the first finger is longer than the second, the digital expansions are larger, there is an outer metatarsal tubercle, and the upper lip is margined with white.

29. *Dendrobates tinctorius*, (Schneid.). Bouleng., Cat. Batr. Ecaud., p. 142.

Hab. Tanti (1890 feet). Two specimens.

30. *Phryniscus lævis*, Gthr., Cat. Batr. Sal., p. 43, pl. iii, fig. A.

Hab. Between Latacunga and Machachi (9000 to 11,500 feet); Panecillo, Quito (10,000 feet); Hacienda of Guachala (9217 feet); and Riobamba (9000 feet). Eleven specimens.

134   *TRAVELS AMONGST THE GREAT ANDES.*

Of this frog Mr. Whymper says, "it is one of the most widely distributed, I think *the* most widely distributed, in the interior of Ecuador. I have seen it almost everywhere from 7000 feet above the sea to 13,500 feet. We could have obtained thousands of specimens."

31. *P. elegans*, Bouleng., Ann. and Mag. N. H. (5), ix, p. 464.

Hab. Tanti (1890 feet). A single ♀ specimen.

Head small, its length nearly one-third that of the body; snout prominent, truncate, not pointed, a little longer than the diameter of the eyeball; canthus rostralis angular; loreal region vertical; rostral nearer the tip of the snout than the eye; interorbital space broader than the upper

PHRYNISCUS ELEGANS, BOULENGER.
TANTI, 1890 FEET.

eyelid. Limbs slender; stretched along the body, the fore limb extends beyond the vent by the length of the fourth finger, the hind limb marks the middle of the eye with the tibiotarsal articulation. Fingers slightly webbed at the base, first very short; toes nearly entirely webbed, the last two phalanges of fourth toe free; inner toe very short, but perfectly distinct; no subarticular, nor carpal, nor metatarsal tubercles. Skin perfectly smooth. Light pinkish grey above, vermiculated with broad black lines; a black streak from the tip of the snout, through the eye, along each side of the body to the groin; lower surfaces white, immaculate, except a few small black spots under the limbs. From snout to vent 34 millim.

32. *Hylodes conspicillatus*, Gthr., Cat. Batr. Sal., p. 92.

Hab. Milligalli (6200 feet). Two specimens, ♂ and young.

33. *H. unistrigatus*, Gthr., Proc. Zool. Soc., 1859, p. 416.

Hab. Machachi (9-10,000 feet); Chillo (9000 feet); Hacienda of Olalla, plain of Tumbaco (8500 feet). Twelve specimens.

34. *H. Whymperi*, Bouleng., Ann. and Mag. N. H. (5), ix, p. 465.

Hab. Pichincha (11-12,000 feet); Valley of Collanes (12,500 feet); Cotocachi (13,000 feet); Tortorillas, Chimborazo (13,200 feet). Six specimens.

Habit of *Hyla arborea*. Tongue oval, entire. Vomerine teeth in two oblique series behind the choanae. Snout rounded, as long as the greatest orbital diameter, with distinct canthus rostralis; nostril nearer the tip of the snout than the eye; interorbital space a little broader than the upper eyelid; tympanum hidden. Fingers moderate, first shorter than second; toes moderate, quite free; disks and subarticular tubercles moderate; two metatarsal tubercles. The hind limb being carried forwards along the body, the tibio-tarsal articulation reaches the angle of the mouth. Skin of upper surface tubercular; on the back the tubercles are confluent into more or less distinct longitudinal lines; belly granulate. Dark olive-brown above (in one specimen with a few light spots); greyish or reddish brown, immaculate or marbled with dark brown, beneath; upper lip whitish.

HYLODES WHYMPERI, BOULENGER.
PICHINCHA, COTOCACHI, ETC.

In the specimen from the valley of Collanes and in that from the mountain Cotocachi the front and hinder sides of the thighs are tinged with magenta red. From snout to vent 27 millim.

H. Whymperi resembles *H. unistrigatus* in general appearance; but the latter has a distinct tympanum, larger digital expansions, the skin smooth, or nearly smooth, above, and a strong fold across the chest.

Besides these three well-characterised *Hylodes* there are eleven very small specimens from Chillo, 9 to 15 millim. long from snout to vent, too young to be properly determined. Upon these Mr. Whymper observes, "This miniature species was first brought to my notice by an English resident; and he assured me that the largest of the specimens represents the full size of the species." This is evidently a mistake, as all the specimens prove to be young. They perhaps belong to a new species; but with the materials before me I will not venture to describe it.

35. *Bufo coeruleostictus*, Gthr., Proc. Zool. Soc., 1859, p. 415.
  Hab. Nanegal (3-4000 feet). One male specimen.

36. *B. marinus*, (L.). Bouleng., Cat. Batr. Ecaud., p. 315.
  Hab. Near the Bridge of Chimbo (900 feet). Two very young specimens.

37. *B. crucifer*, Wied. Bouleng., Cat. Batr. Ecaud., p. 316.
  Hab. Tanti (1890 feet). Two half-grown specimens.

38. *Nototrema marsupiatum*, (Dum. & Bibr.). Erp. Gén., viii, p. 598, pl. xcviii.
  Hab. Machachi (9-10,000 feet); Hacienda of Antisana (13,300 feet); Pedregal (11,600 feet). Numerous examples.

  Mr. Whymper says that the ground colour was bright green. "These frogs were in great numbers in the neighbourhood of Machachi, and in the evening their music was so loud as almost to interfere with hearing when walking out."

APODA.

39. *Coecilia pachynema*, Gthr., Proc. Zool. Soc., 1859, p. 417.
  Hab. Milligalli (6200 feet). One specimen.

40. *C. gracilis*, Shaw. Bouleng., Cat. Batr. Caud., p. 95.
  Hab. Guayaquil. One specimen.

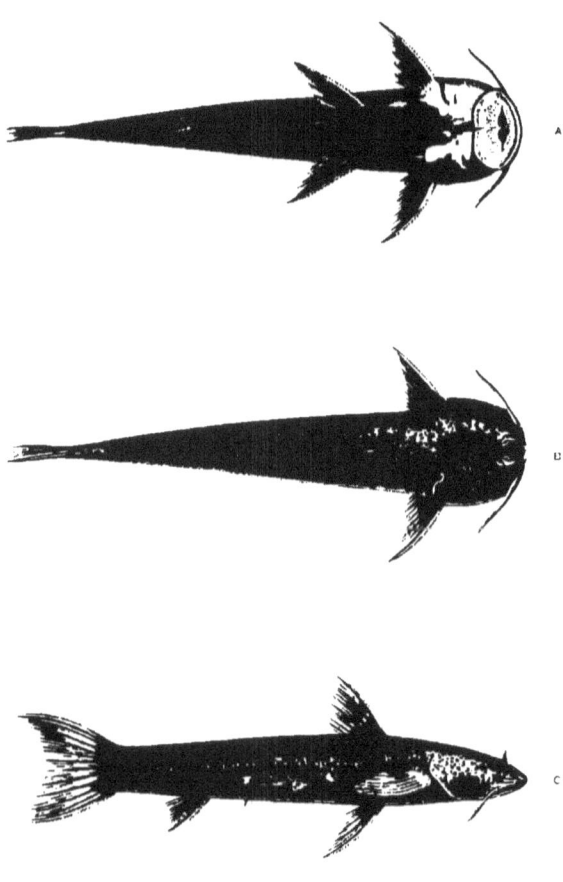

A. VIEW FROM BELOW   B. VIEW FROM ABOVE   C. SIDE VIEW

CYCLOPIUM CYCLOPUM, Humboldt.

By F. DAY, C.I.E., F.L.S., F.Z.S.

I received in Sept. 1883, Mr. Whymper's collection of these small siluroid fishes, in excellent condition, amounting to 51 specimens, which were obtained at Machachi (9500), Chillo (8500), neighbourhood of Cayambe (9-9500), and Riobamba (9000).[1] The streams from the three first-named of these localities flow into the Pacific, and from the last pass into the Amazons.

B. iv., D. $\frac{1}{6}$ | 0-1, P. $\frac{1}{9}$, V. $\frac{1}{4}$, A. $\frac{1}{5\text{-}6}$, C. 13.

Length of head about $4\frac{1}{2}$ to 5, of caudal fin $4\frac{3}{4}$ to $6\frac{2}{3}$ in the total length, —among four specimens from Chillo being as follows:—

Length of specimen 3·6 inches ; of head $4\frac{1}{2}$ of caudal fin $5\frac{1}{4}$ in the total length.
,, ,, 3·0 ,, ,, 5 ,, $6\frac{2}{3}$,, ,,
,, ,, 1·9 ,, ,, $4\frac{2}{3}$,, $4\frac{3}{4}$,, ,,
,, ,, 2·0 ,, ,, $4\frac{1}{2}$,, 5 ,, ,,

The above figures show that a considerable individual variation in proportions exists. The width of the head as a rule equals its length, but varies from $\frac{4}{6}$ to $\frac{9}{9}$ of that extent. *Eyes* small, situated in or slightly behind the middle of the length of the head. *Lips* thick, broad, especially the lower, and evidently used as a sucker ; nostrils—separated by a well-developed flap—situated one-half nearer the snout than the orbit, and a barbel at the angle of the mouth which reaches as far as the orbit. The skin on the upper surface of the head rough. Gill-openings separated by a broad isthmus *Teeth*—in the upper jaw in several rows, the outer of which are flattened and of a club-like form in their outer ends, while the inner rows of teeth have y-like branched outer extremities. In the lower jaw the teeth are larger

---

[1] As a detailed account of the small siluroids of the Andes has been given by Dr. F. W. Putnam (American Naturalist, 1871, p. 694) it will be unnecessary to again go through them at much length. The conclusions there arrived at were formulated from an investigation into a specimen from Quito, Ecuador, and from the descriptions by Dr. Günther of *Arges brachycephalus*, founded on two specimens from the Andes of Western Ecuador, and *Stygogenes Humboldtii*, Günther, described from four specimens up to two inches in length from Madame I. Pfeiffer's collection D. $\frac{1 \cdot 6}{6 \cdot 1}$ | 0-1, A. 6, C. 13, P. 8-10, V. 6.

Putnam suggested that the spine in the second dorsal fin of Humboldt's specimen was overlooked, and also in Dr. Günther's specimen of *Stygogenes*, but more probably it was atrophied.

than in the upper; the outer row being y-shaped, while the two or three inner rows are similar in form but much smaller. *Fins*—the dorsal is situated in the anterior half of the total length, excluding the caudal fin, its first ray ⅔ the length of the head is semi-osseous in its lower portion, while its anterior and lateral surfaces are covered with fine spines directed upwards,— its remaining rays are branched. There is no distinct second dorsal in the larger examples, but in the two smaller ones there is a constriction along the body as if an adipose fin had extended from the first dorsal to the caudal but which becomes lost in the larger specimens. Caudal—emarginate, with the outer rays spinate as in the first dorsal ray, but to a less extent. Pectoral commences under the gill-opening, its outer ray very similar to the first of the dorsal, but stronger and ¼ longer, and extending to above the ventral.

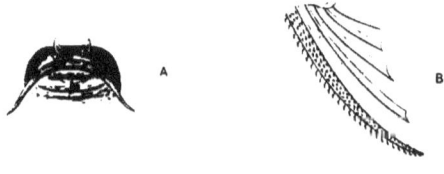

CYCLOPIUM CYCLOPUM, HUMBOLDT.
A, HEAD, FRONT VIEW. B, VENTRAL FIN.

The ventral inserted on a line beneath the third or fourth dorsal ray. It is rather longer than the pectoral, while its outer ray is of a similar character. It does not extend to so far as the anal, which latter has the first ray similar to those of the other fins but shorter. *Skin* rugose, and covered with small, porous orifices. *Lateral-line* distinct, consisting of small circular orifices, and in the last half of the body of small tubes. *Colour*, now of a leaden hue, having a yellowish tinge beneath.

Four specimens were received from Riobamba.

Length of specimen 2·2 inches; of head 4⅞ of caudal fin 7 in the total length.
,, ,, 2·1 ,, ,, 4¼ ,, 7 ,, ,,
,, ,, 1·9 ,, ,, 4¾ ,, 6½ ,, ,,
,, ,, 1·8 ,, ,, 4½ ,, 6 ,, ,,

This form differs from the preceding one in being proportionately a little more elongated and being more or less covered with black markings. The largest has no distinct adipose dorsal, but the same indistinct appearance as observed in some of the other fishes; the one second in size (2·1 inches) has this indistinct adipose fin with a small spine at its posterior termination a little behind a line with the posterior end of the anal fin. The third (1·9 long) has no appearance of an adipose dorsal spine. The smallest (1·8 long) has a distinct spine with four spinate elevations along its upper surface, and which are stronger than the similar elevations in the other fins. This series is very interesting, for it demonstrates that a spine in the adipose fin may be present or absent, and likewise that the spine if present may be smooth or spinate along its upper edge. While these differences alone may be observed in these siluroid fishes, which otherwise entirely resemble one another, they may reasonably be considered to be identical species.

Ten specimens from the neighbourhood of Machachi. One was examined and found to contain 13-16 vertebræ, while the air-bladder was enclosed in bone as in the loaches. It would be tedious to follow out each specimen as all agreed with the before-described form. The five smallest from 0·9 inches to 1·5 inches in length show the adipose dorsal spine with its spinate upper edge. From 1·8 inches in length to 2·4 the long and low adipose fin is still visible, but in another at 2·4 inches it has entirely disappeared. Six other specimens taken from near the same locality and all of small size show the adipose dorsal spine—this series evidently proving that not only may the adipose dorsal spine be atrophied in large examples but also that all trace of the dorsal fin itself may be lost.

Twenty-seven specimens were received from the neighbourhood of Cayambe. The lengths of these fish vary from 0·7 of an inch to 2·7 inches, while most are infested with parasites. Precisely the same characters are shown, but the adipose dorsal fin is not so completely lost in the larger ones as in those from Machachi. This would seem to be due to the examples being rather better prepared. Were these specimens kept in weak spirit doubtless they would take on the appearance of a long, low adipose dorsal fin.

As to breeding the ova are comparatively large, while the male appears to possess an intromittent organ.

My examination of the foregoing series of siluroids from the Andes induces me to coincide with Dr. Putnam's views. 1. That the adipose dorsal spine has not even a specific value. Although all four specimens from Chillo agree with *Arges brachycephalus*, Günther, so likewise do the larger examples with *Stygogenes Humboldtii*, Günther. 2. That the names of the fish have the following priority. *Cyclopium* (Swainson) *cyclopum*, Humboldt : *Pimelodus cyclopum*, Humb., 1803 : *Cyclopium Humboldtii*, Swainson, 1839 : *Arges cyclopum*, Cuv. and Val., 1840 : *Arges brachycephalus*, Günther, 1859 : *Stygogenes Humboldtii*, Günther, 1864 : and *Stygogenes cyclopum*, Günther, 1864.

NOTE ON ROCKS FROM THE ANDES.

BY PROFESSOR T. G. BONNEY, D.Sc., F.R.S.

During his journey in the Ecuadorian Andes, Mr. Whymper collected a large series of rocks, which he has placed in my hands for microscopic examination.[1] The collection has an exceptional interest from the fact that so many of the specimens have been obtained from localities never visited before by travellers, and in not a few cases by any human being, for among them are fragments from the highest crags of several peaks hitherto unascended. Mr. Whymper endeavoured to make his collection thoroughly representative, so that the specimens are in various stages of preservation; but there has always been a fair proportion in as good a condition for examination as could reasonably be expected from localities where there are no quarries and the rainfall is heavy. The number of specimens from the different localities varies. From some mountains, such as Pichincha, Antisana, and Chimborazo, it is large,—from others only two or three specimens have been brought. This has been partly due to the exigencies of travel, and yet more to the fact that there is such a general resemblance among the igneous rocks of the whole district that it was in many cases obviously useless to accumulate specimens which would be lithologically duplicates, differing only in localities.[2]

With the exception of one mountain (Sara-urcu), which is composed of metamorphic rocks, all the great peaks of the Andes ascended by Mr. Whymper consist of one species of volcanic rock—that to which the name *Andesite* has been given. This name, the derivation of which is obvious, was first applied by Von Buch to the rock, as was the name *Andesine* to a species of felspar which it was supposed largely to contain. This felspar is regarded as a member of the plagioclastic group, and contains nearly equal amounts

[1] Descriptions of these have been read before the Royal Society during the year 1884, and the five parts are published in their Proceedings for that year.

[2] The collection included specimens of the highest rocks of the four loftiest of the Ecuadorian Andes, viz. — Chimborazo (20,498 feet), Cotopaxi (19,613 feet), Antisana (19,335 feet), and Cayambe (19,186 feet); and others from the actual summits of Carihuairazo (16,515 feet), Cotocachi (16,301 feet), Sincholagua (16,365 feet), Pichincha (15,918 feet), Corazon (15,871 feet), and Sara-urcu (15,502 feet).

of lime and soda, occupying thus a position intermediate between oligoclase (which has distinctly more soda than lime), and labradorite (which has more lime than soda). The silica percentage also is intermediate between those of the same two felspars. Whether there be such a felspar, capable of distinction as a species, we must leave to mineralogists to decide; and to settle how many are the true species of felspar is a question not easy to solve. The rock *Andesite*, however, is a well-recognised and fairly definite one. The dominant felspar is always plagioclastic, and commonly, if felspar be present both in distinct crystals and in microliths, the former contain the lower percentage of silica and approximate to labradorite, the latter contain the higher and come nearer to oligoclase. Chemically the rock is distinguished by the predominance of soda among its alkaline bases. The percentage of silica may fall rather below 60, in the more basic varieties, or rise up to more than 70 in the most acid varieties (often called Dacites) in which free quartz is usually present. In the quartzless group, however, the silica percentage is generally not more than 65.[1] Lithologically the rock consists of a glassy base (often crowded with microlithic products) in which are usually scattered larger crystals of felspar, of a pyroxenic mineral, and of iron-oxide. The pyroxenic constituent may be one or more of the following:—augite, hornblende, hypersthene;[2] occasionally there is a little biotite.

The great majority of the specimens collected by Mr. Whymper belong to the quartzless division of the *Andesites*, only an outlying part of one mountain (Antisana) having furnished varieties which have a silica percentage exceeding 70. The latter are pitchstones of ordinary aspect, being probably very vitreous members of the Dacite group. The most basic forms, represented by the nearly black, subvitreous rock, which but for a peculiar resinous aspect and rather too low specific gravity might readily be classed as a basalt (the melaphyre and the pitchstone porphyrite of some authors), also seem to be rare; the majority of the specimens varying in colour from light greyish or yellowish or reddish to duller tints of grey or red, which in only a few cases approach black.

I give below a table of the volcanic products of each mountain, enumerating them from north to south.[3]

[1] Numerous analyses have been published. Several will be found in Zirkel, Microscopical Petrology (Report of the Geol. Expl. of the 40th Parallel, U.S.A.).

[2] It has been doubted whether this be only a dimorphic (rhombic) form of augite, or true hypersthene, but recent investigations show that its analysis warrants the reference to the latter mineral.

[3] The following explanation may be useful. If the observed pyroxenic constituent is wholly, or almost wholly, augite, the rock is called an *augite-andesite*; if hornblende, a *hornblende-andesite*; and when there is a marked amount of hypersthene the epithet *hypersthenifcrous* is added, and so on.

WESTERN ANDES.

COTOCACHI. *Hyperstheniferous augite-andesite.*

PICHINCHA. Highest peak, chiefly *hornblende - andesites.* Second peak (RUCU - PICHINCHA) *hyperstheniferous augite-andesites.*

CORAZON. *Augite-andesites,* usually *hyperstheniferous* ; in one case (summit ridge) rather rich in this mineral.

ILLINIZA. Few specimens. The best preserved a *hornblendic augite-andesite,* with some *mica* and *hypersthene.* The summit rock a *micaceous andesite.*

CARIHUAIRAZO. Summit rock *augite-andesite,* perhaps *hyperstheniferous.*

CHIMBORAZO. *Augite - andesites,* generally *hyperstheniferous;* sometimes almost entitled to be called a *hyperstheneandesite.* Occasionally *hornblende* is also present.

EASTERN ANDES.

CAYAMBE. Generally *augite-andesite,* but with considerable minor variations of the pyroxenic constituent ; best general term *hornblendic augite-andesite :* one a *mica-andesite.*

ANTISANA. The main mass of the mountain *augite-andesites,* sometimes *hyperstheniferous.* From an outlying part of the mountain come some pitchstones, probably very glassy *dacite.*

SINCHOLAGUA. *Augite - andesite,* probably *hyperstheniferous.*

COTOPAXI. *Hyperstheniferous augite-andesite.*

ALTAR. *Augite-andesites,* probably in general *hyperstheniferous.* Some of the rocks seem to come nearer to basalt in their composition than is usual in these specimens from the Ecuadorian Andes.

Sara-urcu, the only non-volcanic peak ascended by Mr. Whymper, lies to the east, rather to the south-east of Cayambe. Among its rocks are varieties of mica-schists and fine-grained gneisses, and a specimen of 'spotted schist,' consisting chiefly of small crystals of mica, and a mineral or minerals belonging to the non-alkaline alumina silicates. I am a little doubtful whether the felspar-like mineral in the so-called gneisses may not be rather kyanite or some kindred mineral. So far as I can judge from the lithological characters of the rocks I should regard them as Archaean, and compare them with the definitely bedded schists, which usually belong to the middle or upper part of that great series.

Mr. Whymper nowhere saw crystalline igneous rock in the mountain region of the Andes. From a boulder in a stream at the western margin he brought a specimen of tonalite (or quartz-diorite) ; and from the district east of that visited by him he secured a specimen of granite.

He brought back a considerable series of specimens of volcanic dusts from

Cotopaxi; one of them having been ejected by that volcano and collected upon the slopes of Chimborazo under the exceptionally interesting circumstances detailed in his narrative. Of these it may suffice to say that they consist of minute lapilli, chips of volcanic glass, and fragments of crystals of plagioclastic felspar, augite, and hypersthene (I have noticed one or two very characteristic crystals of the last). They are in short the materials which one might expect to obtain from the explosive destruction of lavas such as occur on the cone of the mountain. The glass is moderately solid, differing thus (for example) from the very vesicular ejectments of Krakatoa—the constant 'puffing' of Cotopaxi probably preventing the accumulations of such a large amount of steam as to permeate completely the more or less molten matter in the pipe of the crater. The dust which fell on Chimborazo, after an aerial voyage of full 65 miles, consisted of fragments varying from about four-thousandths of an inch downwards, very many ranging from about one-thousandth to three-thousandths. Specimens of a coarser ash of a pumiceous character have also been brought, which no doubt also belongs to the same group as the more solid rocks.

INDEX TO GENERA.

BATRACHIA.

Bufo, 136.
Cœcilia, 136.
Dendrobates, 133.
Hylodes, 135.
Nototrema, 136.
Phryniscus, 133-4.
Prostherapis, 132-3.

COLEOPTERA.

Achryson, 37.
Ægithus, 56-7.
Agonum, 16.
Alethaxius, 83.
Amathynetes, 61, 70-1.
Amphideritus, 61, 68.
Ananca, 44.
Anchomenus, 12, 13.
Anchonus, 61, 72-4.
Ancognatha, 30.
Ancylonycha, 28.
Anisotarsus, 4, 7, 8.
Anomala, 28.
Anthocomus, 52.
Antichira, 29.
Aphthona, 85-6.
Apion, 61, 78-9.
Arescus, 54-5.
Artematopus, 45-6.
Asiopus, 43.
Astæna, 26.
Astylus, 52-3.
Athöus, 44-5.
Attalus, 52.
Barothrus, 30-1.
Baryxenus, 32-3.
Bembidium, 4, 22-4.
Brachysphœnus, 56.

Brenthus, 80-1.
Calandra, 80.
Calathus, 16.
Calligrapha, 83.
Calopteron, 46.
Carabus, 3.
Carneades, 39.
Cercometes, 58.
Cercus, 59.
Ceroglossus, 3.
Chauliognathus, 51.
Chelymorpha, 55.
Chlorida, 37.
Chlorota, 29.
Chrysomela, 84.
Cladodes, 47-8.
Clavipalpus, 27.
Clœotus, 26.
Coccinella, 57.
Colaspis, 82.
Colpodes, 4, 13-22.
Colymbetes, 40.
Compsus, 61, 63-4.
Coptocycla, 55-6.
Cossonus, 61, 80.
Cratomorphus, 48-9.
Cryptobium, 41.
Cyclocephala, 31.
Cycloneda, 57-8.
Cyllene, 38.
Dasytes, 53.
Daulis, 58.
Diabrotica, 87-8.
Dibolia, 86.
Diphaulaca, 86.
Discodon, 49.
Doryphora, 83-4.
Dyscinetus, 32.
Eburia, 37.

Enema, 33.
Epitragus, 42.
Epitrix, 84.
Epuræa, 60.
Eriopis, 57.
Erirhinoides, 61, 76.
Erirhinus, 61, 76-7.
Erycus, 77.
Estenorrhinus, 80.
Eurysthea, 37-8.
Exorides, 65.
Galerita, 24.
Golofa, 34.
Gymnetis, 34-5.
Haltica, 85.
Hammoderus, 39.
Haplamaurus, 54.
Haptoderus, 12.
Helicorrhynchus, 61-2.
Heterogomphus, 33.
Hilipus, 61, 74-6.
Homoiotelus, 56.
Hypsioma, 39.
Lasiocala, 29.
Leucopelæa, 30.
Listroderes, 61, 69-71.
Listrus, 53-4.
Longitarsus, 84.
Luperosoma, 87-8.
Luperus, 87.
Lyterius, 78.
Macrops, 4, 61, 72.
Mallodon, 37.
Malthesis, 52.
Megaceras, 32.
Megilla, 57.
Meloe, 43-4.
Metamasius, 79.
Migadops, 3.

INDEX TO GENERA.

Morphoides, 56.
Naupactus, 61, 65-8.
Nebria, 3.
Neda, 57.
Neleus, 36.
Nitidula, 60.
Noda, 82.
Nyctobates, 42.
Otidocephalus, 61, 77-8.
Omaseus, 10.
Oncideres, 39.
Ontherus, 25.
Otiorrhynchus, 61-2.
Pandeletius, 61-3.
Parandra, 36.
Passalus, 35.
Pelmatellus, 4, 8, 9, 10.
Percus, 13.
Pertinax, 36.
Phanæus, 26.
Phengodes, 49.
Philonthus, 4, 40, 41.
Phoroneus, 35-6.
Photinus, 48.
Physonota, 55.
Pinotus, 26.
Plateros, 46-7.
Platycelia, 4, 29, 30.
Plectonotum, 51-2.
Pleuroneces, 59-60.
Pœderus, 41-2.
Polipochila, 10.
Polydacris, 63.
Praepodes, 65.
Praogolofa, 34.
Prionocalus, 36-7.
Pristoscelis, 54.
Pseudoxychila, 7.
Pterostichus, 10-12.
Ptilodactyla, 46.
Pyrophorus, 45.
Rhantus, 40.
Sciopithes, 62.
Scymnus, 58.
Semiotus, 45.
Silis, 51-2.
Silpha, 4, 40.
Sitones, 66.
Sphenophorus, 61, 79.

Stenochrates, 32.
Sterculia, 41.
Stratægus, 34.
Strongylium, 42.
Tæniotes, 39.
Telephorus, 49.
Tenebrio, 42.
Thyridium, 29.
Trachyderes, 6, 38.
Trechus, 22.
Trichaltica, 85.
Trigonogenius, 4, 51.
Trogosita, 60.
Trox, 26.
Uroxys, 24.
Xenismus, 50, 51.

CRUSTACEA.

Allorchestes, 126.
Amphilhoë, 126.
Aratus, 123.
Boscia, 121.
Cancer, 124.
Gammarus, 126.
Gelasimus, 123.
Hyale, 126.
Hyalella, 125-7.
Macrobrachium, 124.
Metoponorthus, 125.
Orchestia, 126.
Palæmon, 124.
Philoscia, 125.
Porcellio, 125.
Pseudothelphusa, 121-2.
Sesarma, 123.
Squilla, 124.

HYMENOPTERA.

Atta, 95.
Camponotus, 89-91.
Ectatomma, 91.
Holcoponera, 92-3.
Odontomachus, 93.
Pachycondyla, 91.
Pheidole, 93-5.
Ponera, 91.
Pseudomyrma, 95.

LEPIDOPTERA.

Achlyodes, 110.
Acræa, 96, 99.
Actinote, 99.
Adelpha, 103.
Aganisthos, 103.
Agraulis, 96, 100.
Amphirene, 102.
Anæa, 103.
Anartia, 101.
Ancyloxypha, 96, 110.
Argynnis, 4.
Caligo, 99.
Callicore, 102.
Callidryas, 107.
Catagramma, 102.
Ceratinia, 97.
Chionobas, 4.
Cœa, 103.
Colenis, 100.
Colias, 4, 96, 107-8.
Corades, 99.
Danais, 96.
Didonis, 102.
Emesis, 104.
Epinephele, 4.
Erebia, 4.
Eresia, 100-1.
Euptoieta, 96, 100.
Eurema, 101.
Euterpe, 105.
Heliconius, 97, 100.
Hesperia, 109.
Hesperocharis, 105.
Hyposcada, 97.
Ithomia, 97-8.
Ituna, 97.
Junonia, 96, 101.
Lasiophila, 98.
Leptalis, 105.
Lycæna, 96, 104, 105.
Lymanopoda, 96, 98.
Marpesia, 103.
Mechanitis, 97.
Meganostoma, 96, 107.
Mesosemia, 104.
Morpho, 99.
Mylothris, 107.

INDEX TO GENERA.

MYSCELIA, 101.
NYMPHALIS, 103.
OPSIPHANES, 99.
PAMPHILA, 96, 110.
PAPILIO, 96, 99, 100, 103, 107-9, 110.
PARNASSIUS, 4.
PAVONIA, 99.
PEDALIODES, 96, 98.
PERISAMA, 102.
PHYCIODES, 101.
PIERIS, 4, 96, 105-7.
PREPONA, 103.
PRONOPHILA, 98.
PROTEIDES, 109.
PYRAMEIS, 96, 101.
PYRGUS, 110.
PYRRHOPYGA, 109.
PYTHONIDES, 110.
SISEME, 104.
STEROMA, 96, 98.
TERIAS, 96, 106.
THYMELE, 109.
TIMETES, 102-3.

PISCES.

ARGES, 137, 139.
CYCLOPIUM, 137, 139.
PIMELODUM, 139.
STYGOGENES, 137, 139.

REPTILIA.

AMEIVA, 129.
AMPHISBÆNA, 130.
ANOLIS, 128-9.
BOA, 130.
BOTHROPS, 132.
CINOSTERNUM, 128.
CORONELLA, 130-1.
ECPLEOPUS, 129.
ELAPS, 132.
GONATODES, 128.
GYMNODACTYLUS, 128.
HERPETODRYAS, 132.
HIMANTODES, 132.
HOLOTROPIS, 129.
HOMALOCRANION, 130.
LEPTOGNATHUS, 130, 132.
LIOCEPHALUS, 129.
LIOPHIS, 130, 131, 132.
OPHEOMORPHUS, 132.
OXYRHOPUS, 132.
PROCTOPORUS, 130.

RHYNCHOTA.

ACANTHIA, 118.
ACUTALIS, 120.
ANEURUS, 115.
AROCERA, 112.
CALOCORIS, 114.
CARINETA, 119.
CIMEX, 111, 115.
CINYPHUS, 114.
CONORHINUS, 115.
DIONYZA, 114.
DIPLONYCHUS, 118.
EMESA, 117.
GEOTOMUS, 111.
HARMOSTES, 113.
HERANICE, 120.
LYDE, 114.
LYGUS, 114.
MARGUS, 113.
NEOMIRIS, 113.
NEZARA, 112.
NYSIUS, 113.
PENTATOMA, 112.
PIEZODORUS, 112.
PNOHIRMUS, 117.
PRIONOTUS, 115.
REDUVIUS, 115.
RHAPHIGASTER, 112.
SEPHINA, 112.
SPHENORHINA, 119.
STENOPODA, 116.
TETTIGONIA, 120.
THELIA, 120.
THYANTA, 111.
ZAITHA, 118.
ZAMMARA, 118.

THE END.

www.ingramcontent.com/pod-product-compliance
Lightning Source LLC
Chambersburg PA
CBHW032228230426
43666CB00033B/1639